A TALES OF TECHNOFICTION BOOK

VISIONS OF 2051

MORE ON THE RISING CYBER MUSES

ROGER BOURKE WHITE JR.

authorHOUSE®

AuthorHouse™
1663 Liberty Drive
Bloomington, IN 47403
www.authorhouse.com
Phone: 1 (800) 839-8640

© 2018 ROGER BOURKE WHITE JR. All rights reserved.

No part of this book may be reproduced, stored in a retrieval system, or transmitted by any means without the written permission of the author.

Published by AuthorHouse   06/15/2018

ISBN: 978-1-5462-4706-7 (sc)
ISBN: 978-1-5462-4705-0 (e)

Library of Congress Control Number: 2018907027

Print information available on the last page.

Any people depicted in stock imagery provided by Getty Images are models, and such images are being used for illustrative purposes only.
Certain stock imagery © Getty Images.

This book is printed on acid-free paper.

Because of the dynamic nature of the Internet, any web addresses or links contained in this book may have changed since publication and may no longer be valid. The views expressed in this work are solely those of the author and do not necessarily reflect the views of the publisher, and the publisher hereby disclaims any responsibility for them.

Contents

Overview vii

The Essays
Gene Editing 1
Food Production 7
The Challenge of Space Commerce 10
Warfare 14
Electioneering in VR 16
Robot Companions 18
The Evolution of Wearables 20
Personal Performance Enhancement and Wearables 25
Mixing Romance and Wearables 30
Desire and Wearables 36
Us versus Them Thinking and Wearables 39
Updating hardware and software 42
Three Kinds of Money 45
Education: Necessity Style and Ambitious Style 52
Health Care 55
Coming of Age Rituals 59
Baby Clubs 62
The Baby Club Lifestyle 65
Cosplay 71
Ambitious Class Mind-Tripping 73
Achieving new states of consciousness 76
Recreational Mind Altering 78
Substance Abuse 81

The Paralympics Crisis 83
Gaming the System 86
Dissent and Social Shaming 88
Preppers 91
What will thinking be like in TES? 97
Leisure Time in TES..103
Enfranchisement in TES 107
Social Classes in TES 111
Immigration in TES 116
The "paycheck to paycheck" Lifestyle in TES.......................... 119
The Shopping Experience in TES 127
Charity and TES.. 129
Politicians in TES 130
Social Shaming in TES 135

The Stories
The Unexpected Hero 141
Jeanie The Gene Editor 143
Wise Old Man -- JC version...145
Alien Girlfriend...149
He's The One? 160
Party Hearty or Study Hearty?...171
The Princess, The Fly, and City Hall 180
Hard Times 02... 188
It's Time... 207
Megan's Party Time 226

Epilog: What's coming after the 2050's?................................. 241

Overview

Welcome to the world of the 2050's... again.

This book is a continuation of the forecasting I have done in my book Visions of 2050. I will be talking more about driverless cars and cyber muses, and I will be discussing lots of new topics as well such as third generation wearables, new kinds of money, and genetic editing that is going to get much easier and cheaper.

On the social side I will be talking about how all these new technologies and inventions are going to change how we humans relate to each other and to the world around us. With cyber handling the heavy lifting in providing our material prosperity humans are going to find other activities besides jobs (in the 2010's meaning of that word) to keep them busy and excited about accomplishing things.

And I will be mixing in lots of short stories that give a nice human twist to describing what these changes will be like.

All-in-all, if you are interested in how we will be living in the 2050's you should find this an interesting read.

Book Structure

The structure of this book is to have two styles of writing.

The first style is essays discussing various technologies and social implications that I forecast are coming and the significance and surprises that are going to come with them. How are these changes going to alter how we live and what we think about? As an example, how is our thinking about cars going to change as driverless cars become ubiquitous? These are technology forecasting and social forecasting essays.

The second style is short stories. These are entertaining stories that delve further into how our lives are going to be changed by these upcoming technologies and social changes.

The goal of these styles is to look at upcoming changes from different points of view. This can give you, the reader, better insights into what is coming.

One change in a starting premise

One starting premise for these Vision books is that technology is going to continue to improve, but basic human thinking -- instincts and emotions -- is not going to change much. This relation has been a given since the beginning of the Industrial Revolution back in the 1770's. This interacting between the new possibilities in changing our material world that technology brings and the constants of instinctive and emotional human thinking is what has produced our fascinating human history since the Industrial Revolution began. The Vision books are forecasting how this fascinating history will continue to evolve.

But in this book there will be one change to the above premise. In this book technology will begin to change human emotional and instinctive thinking. In this book I will be talking about third generation wearables and those will have the ability to modify emotions, and modifying emotions can influence instinctive thinking. This is yet another whole new ball game that technology will be bringing to human social relations.

Yes, the 2050's are going to be surprising times in many ways.

THE ESSAYS

Technology Revolutions

This first section of the book is about the technology revolutions that will be coming to us over the next forty years. Technology change has been accelerating since the start of the Industrial Revolution and that is going to continue. Lots of technology changes are coming and this book will be about a handful of high profile ones.

Commodity Uses and Surprise Uses

Most technologies start as a hot idea in an inventor's mind. That hot idea has to become something tangible if it is going to shake up the world. But the first step in this process is usually quite prosaic: this hot idea is first going to be used to replace an existing product or service, but do it faster, better and cheaper. This first use will be ho-hum, not world shaking, but it will pay for the basic research and development that will turn the hot idea into something tangible. I call this the "commodity use" for the hot idea. Then, if this hot idea is truly world-shaking, people using these first implementations will notice that the tool can get used for something much more interesting, as in, "Oh… you can do that with it too? How neat!" I call this the "surprise use" and this is the big step on the road to becoming world shaking. This is what I'm talking about when I refer to commodity uses and surprise uses in the essays that follow.

Gene Editing

Introduction

Genetic editing is going to undergo its version of Moore's Law. Over the next few decades it is going to get faster, cheaper and easier. The tools are going to get easier to use, what can be changed with gene editing is going to get wider in scope, and the kinds of people who will become gene editors will widen dramatically. By the 2050's it is not going your grandpa's kind of genetic editing world. This evolution is going to be much like that which computers experienced as they evolved from mainframes into personal computers.

Gene editing basics

Genes are information. They tell a living organism, a cell, how to make proteins. They are located on a chemical compound called DNA which exists within the cell. When a cell divides into two cells, the DNA replicates itself and half goes into each part of the now divided cell.

In simple cells such as bacteria (also called prokaryotes) the DNA mixes in with all the other cell components. It is also comparatively simple in structure. In complex cells, those that make up plants, animals and fungi, much of the DNA is in a special structure called the nucleus -- much but not all. These more complex organisms are called eukaryotes and their DNA makes a much wider variety of cells and those cells are much more specialized in their functions. Think of the difference between a heart cell and a brain cell in your own body. They both contain the same DNA, but different proteins are being made from that DNA. The difference in proteins is what makes the cells act differently.

Gene editing consists of modifying the DNA that genes are composed of so that different proteins are made, or the timing of when to make the proteins is changed. When a gene makes a protein it is said to be expressing itself. Genetic editing changes how and when genes express themselves in the cell they are living in.

Gene editing in the 2010's

Gene editing in the 2010's is still very much a pioneering technology. It is expensive, the tools are limited in the changes they can make to DNA, and as a result so are the accomplishments. But lots of progress has been made compared to what was available in the 2000's and it looks like even more progress is coming quickly. The latest hot item is CRISPR/Cas9 technology. The tools for genetic editing are getting more numerous and their costs are coming down. This trend is what makes genetic editing much like computing and why Moore's Law is likely to apply.

The 18 Oct 17 Reason article, DIY Biohackers Are Editing Genes in Garages and Kitchens *With the latest breakthroughs in the life sciences, who needs a lab or degree?* by Alexis Garcia & Justin Monticello, talks about the state of DIY biohacking in 2017.

From the article, "These do-it-yourself biologists say the democratization of science has given them the freedom to do work on projects that are often ignored by larger institutions. They're using gene editing technologies like CRISPR to create personalized treatments for those suffering from rare diseases or cancer, reverse engineering pharmaceuticals like Epi-Pens so people can make their own medicine at home, and even creating glow in the dark beer."

My feeling is that the article authors are overemphasizing the human treatment aspects -- this makes sense if you want to make the article sensational. I'd like to hear more about what these people are doing with the simpler organisms, like the glowing dark beer example.

And here is a surprise new ability: adding new codons to DNA and having bacteria use them to put new amino acids into proteins. Wow! Human intervention into a natural process at its finest. This 29

Nov 17 Economist article, A bacterium that can read man-made DNA *Biologists expand life's alphabet to include two new letters*, describes this new process.

From the article, "In a paper published this week in Nature, Dr. Romesberg and his colleagues go a step further, by describing how they have coaxed their bacterium into making proteins containing amino acids that are not found in nature. Each unnatural amino acid to be inserted is represented by a novel codon that includes one of the team's synthetic bases. In other words, their bacterium can quite happily read an entirely new, human-created extension to the standard genetic code, and use the instructions to produce proteins that no organism naturally makes. The hope is that one day this method could be used to make new drugs, polymers or catalysts."

This is also an example of being more innovative on prokaryotes, not humans.

The 7 Dec 17 Economist article, Taking DNA sequencing into the field *With small, portable devices that plug into laptops,* describes how dramatically gene sequencing and editing equipment prices have come down over the 2000's and 2010's.

From the article, "DEVICES for analysing DNA used to be big, clunky and not very good. Hundreds were required for the initial sequencing of the human genome, a project that started in the late 1990s and took over a decade to complete at a cost of at least $500m. Since then, sequencing a human genome has become a routine process; prices have fallen to below $1,000. Although the machines that do the job have got better and more compact, they still cost several hundred thousand dollars. Various groups are trying to make them smaller and cheaper.

The first device small enough to put in your pocket is already on the market. It comes from Oxford Nanopore, a maker of DNA-sequencing equipment based in the eponymous English city. It is about the size of a chunky mobile phone. Although the machine is swathed in patents, other miniature devices are bound to follow in time."

What is coming technologically

What is coming technologically is a lot more variety in what gene editing will be used for. As the processes get simpler, cheaper and more diverse, more kinds of projects will be attempted.

Bacteria/prokaryote editing will be used to make lots of new kinds of chemical compounds. Bacteria cell structure and DNA are comparatively simple, which means they are easier to dramatically modify. They are also better adapted to trying riskier experiments with -- few people care if a bacterium dies in an unnatural way.

Editing on well-known plants and animals, such as crop plants and animals, will be done to produce better varieties of these species. This will be like the breeding that has been done throughout history, but faster, better, cheaper, and smarter -- the breeder won't have to wait for Mother Nature to come up with the mutation they are looking for.

Editing on humans will be similar but it will be subjected to what I call The Curse of Being Important. This means lots of people will have strong opinions on what are good and bad changes, so experimenting and progress will be much slower than what takes place in the world of bacterial experimenting. This is the same phenomenon that we are experiencing in drug research in the 2010's.

What is coming socially

Just as the evolution from mainframe computers to personal computers brought dramatic change in who was using computers and what they got used for, the coming evolution in gene editing technologies is going to change who is doing gene editing and what it gets used for.

The change is going to be towards a more diverse collection of editors and towards a more diverse collection of goals for these editing projects. Think of the difference between Department of Defense employees calculating artillery trajectory tables (one of the first uses for the first computers) and hobbyists designing computer games.

If this pattern holds for gene editing, the number of people involved will grow enormously over the coming decades, and their backgrounds will be quite diverse. This means there will be constant surprises in what kinds of projects get worked on and what gets developed.

One difference between computer evolution and gene editing evolution is the influence of The Curse of Being Important. There was some worry about the consequences of what computers would come up with, there will be more worry about the consequences of what gene editing will come up with.

This concern can shape the evolution in one of two ways.

- If the concern is big and the expense stays high, gene editing will evolve as the health and nuclear industries have. It will remain the province of big businesses and big government organizations. And it will evolve slowly.
- If the concern stays modest and the expense drops dramatically, gene editing will evolve more like personal computing has. It will be decentralized and lots of smaller groups will be at the center of the progress made. It will be more like how the internet has evolved. There will be lots of people expressing worries, but there will be lots of surprising progress happening in spite of those worries.

Talent Agencies

An example of a surprise social institution is gene editor talent agencies. Gene editing is going to remain a complex activity. The more gene editing becomes a skill a person can develop, and the more that skill can become one of personal expression, the more gene editing becomes like the entertainment industry. If it becomes like the entertainment industry, there will be talent agents promoting the virtues of the gene editors they represent.

Gene editing may be brand new, but promoting people's skill at an activity is as old as humanity.

Conclusion

The future of gene editing is going to be surprising. There is the potential for the tools which are used in it to grow in diversity and become much faster, better and cheaper. How much progress can be made along this avenue is currently unknown.

How much progress is made, how diverse the progress is, will depend on two things:

- The tools which can be developed.
- The social concerns about the threats these tools pose.

The progress promises to be widespread. What is developed from editing bacterial genes is going to be worlds apart from what is developed from editing plant and animal genes, and human genes.

Lots of surprises are coming both technologically and socially.

Further Reading

- Wikipedia: Moore's Law - https://en.wikipedia.org/wiki/Moore%27s_law
- Wikipedia: CRISPR - https://en.wikipedia.org/wiki/CRISPR
- DIY Biohackers Are Editing Genes in Garages and Kitchens by Alexis Garcia & Justin Monticello, October 18, 2017 - http://reason.com/reasontv/2017/10/18/diy-biohackers-editing-genes-garages
- A bacterium that can read man-made DNA - Synthetic biology, Nov 29th 2017 - https://www.economist.com/science-and-technology/2017/11/29/a-bacterium-that-can-read-man-made-dna
- Taking DNA sequencing into the field, Dec 7th 2017 - https://www.economist.com/science-and-technology/2017/12/07/taking-dna-sequencing-into-the-field

Food Production

Introduction

How we gather and prepare food is an activity that predates mankind. This means there is a lot of instinctive thinking involved. It is also an activity which has changed dramatically and constantly through the historical centuries -- mankind has been able to successfully apply a lot of analytic thinking to this activity, and he will be able to apply even more in the upcoming decades.

This means there is a complex mix of thinking and activities going on here, and there is lots of change coming in both the thinking and the activities as we evolve into the 2050's.

History

In the Stone Age humans picked out their food from the plants and animals that lived and died naturally around them. Getting food this way is a dicey business filled with lots of uncertainties. This is why the instinct to pay attention to food gathering, quality and preparation is such a strong one in humans of today. Dicey... but on the whole it worked, and humans thrived.

The revolution of the Agriculture Age was to first have the people of a community do some picking and choosing of which plants and animals would live around them, then get deeply into modifying their surroundings so that what they wanted to live thrived mightily -- as in, farming. This increased the quantity and quality of what could be harvested and reduced the uncertainty, a bit, but uncertainty still loomed large in the food acquiring activity. This is why a lot of early religion centered on rituals to make the upcoming harvest a bountiful one.

As the Agricultural Age evolved humans got steadily better at farming, an activity which covers many aspects of making the right foods thrive in ways that create a bountiful harvest for people.

The Industrial Revolution added lots of tools and knowledge to the food acquiring process. This made food producing and preparing faster, cheaper and a whole lot more reliable. This lowering of cost and increasing in variety and certainty is what we are experiencing in the 2010's. For almost everyone in developed communities a regular food supply is a given, and it is a community horror when this isn't true.

So... what's coming next? What will food production be like in the 2050's -- in the Age of Cyber?

What's coming

Traditional farming techniques, such as growing plants on big fields, are going to get even more effective and efficient. Lots of new technologies are going to help with this, two high profile examples being pervasive surveillance and genetic engineering.

These traditional forms will be supplemented by lots of new technologies and techniques. Some will produce food even faster, better and cheaper, and some will cater to human dilettante tastes in this activity. It is not new for many people to want to get up close and personal with their food producing -- suburban gardens are an example of this.

An example of a new way of catering to dilettante tastes will be neighborhood greenhouses. These will be places where people can get up close and personal with the plants that are creating their foods. These people will lavish attention on their food darlings (think pets) so the greenhouses will be filled with technology and loving attention as well as plants.

More exotic food producing techniques will be things such as vat-grown and 3D printed foods. What will be growing in the vats are single-celled organisms such as yeast and bacteria which have been genetically modified to grow into forms which can be transformed into food-resembling materials that seem familiar to the food consumers -- vat-grown hamburgers.

Some will object that this is GMO food, but to my surprise the vat-grown meats have some vegan supporters. They cheer this technology on because it means fewer animals are slaughtered.

What will people care about?

Because food is subject to The Curse of Being Important some of these technologies generate lots of strong opinions. What matters and what doesn't has a lot of capriciousness in it. In the above examples people don't care much if crops are being watched, but they care a lot if they are being genetically altered.

Consistency and reliability will no longer be issues -- these will be taken for granted. But because of the strong instincts surrounding food other issues will become important in their place. A 2010's example of this is people being concerned about how their food was grown, as in, free-range chicken. These kinds of issues will remain important in the 2050's, but the particular topics of concern will be faddish and keep changing -- there will be lots of urban legend mixed in.

Conclusion

The food production and processing revolutions are far from finished. Food production is going to get even faster, cheaper and more reliable, and the technologies used are going to become even more exotic.

Mixed in with this technology revolution is going to be lots of instinctive thinking -- worry about finding good food predates even the human species.

The result is going to be the mixing in of a lot of dilettante interest with all the high technology. Food producing and preparing is going to remain a complex art and one which people are going to stay very much interested in.

The Challenge of Space Commerce

Introduction

Mankind has sought to fly among the stars since prehistoric times. In many cultures their ancient founding gods descended from the stars to come live on earth.

As mankind's technology has improved over the centuries the prospect of returning the favor and visiting the stars has become more and more likely. A "giant leap for mankind" came in 1969 when humans reached the Moon and walked on its surface.

But the Moon is not a star. It is not even a planet. How much further along on this dream will we proceed over the next century? That is the topic of this essay.

The driving forces for exploring: hobby and commerce

There are two styles of thinking that foster exploring: hobby-style curiosity and money-making commerce.

Hobby Style

The hobby style thinking is the curiosity one feels about what lies over the next hill? Should I take the time and effort to go climb that hill and find out? The time and effort can be small -- a day's hike -- or it can get large and elaborate -- mounting an expedition to reach the South Pole. After the Renaissance took place this curiosity has often been mixed with science -- the expedition becomes billed as a science expedition. Reaching the Moon and the South Pole are classic examples of this. But at the root of this style of exploring is the emotion of curiosity.

Commerce Style

Commerce style exploring is based on money making, as in, "Is there profit to be made from taking the trip to that distant land? If so, I'm in!" The classic example of this is Europeans exploring the Americas after Christopher Columbus and Amerigo Vespucci discovered this "new world", and their followers then demonstrating that there were riches to be made taking goods, people and ideas across the Atlantic.

One big difference between hobby style and commerce style is the magnitude of the activity. Hobby style remains small scale while commerce style grows to something enormous in comparison.

What's coming for space?

The hobby style of exploring space has begun. It started with Sputnik back in 1957. The commercial style started shortly thereafter with the launching of communications satellites in the 1960's. Now thousands of flights to space have happened pursuing both hobby and commerce goals. The difference between hobby and commerce style is as evident in space as it is on earth. The commercial activities such as satellite communication, GPS and satellites monitoring what is happening on earth have produced lots and lots of space launches and lots and lots of satellites orbiting earth. The hobby activities, such as putting people on the Moon and the Mars Rover on Mars, remain much smaller in magnitude, but much larger in people's imaginations.

As of now, this difference between commercial activity in earth orbit and hobby activity beyond earth orbit looks likely to continue indefinitely. At this point I can't foresee anything material that will be profitable to move from the various bodies orbiting in the solar system to Earth -- no equivalent to gold, silver and rum coming to Europe from the Americas -- and there is only one thing I can see that will be profitable to move from Earth to any of these bodies orbiting the sun, and that is tourists. (Interstellar is even more so because the costs and journey times are a thousand-fold those of voyages around the solar system.)

So, interplanetary commerce will be based on space tourism. How big an industry that will become is difficult to predict. It will depend a lot on how pleasant and exotic the experience is, as in, how different will a real space vacation be from a VR space vacation experience?

Space Tourism

What will be coming in space tourism?

What I see is an evolution of cruise ship-style tourism happening in space.

One example of this I see coming is a domed city on the Moon, it will be popular just because it is so science fiction. It will feel like Tomorrowland did at Disneyland, and that will be the whole purpose for having it. Oh, there will be talk about supporting science research, and some will be happening, but that is because that is part of the Tomorrowland theme.

The necessity folk will be protesting for "rights" to go there, as they have rights to go to Disneyland. Waiting in line to get such a right will feel OK, just like waiting in line feels OK at Disneyland in the 2010's.

The domed city on the moon is easy to predict. Harder to figure out is what activities other places in the solar system will support. What will Mars support? What will Venus support? What will asteroids support? Asteroids will be the easiest to visit because of their low gravity wells. Venus the toughest because it will be a floating city, not a structure on the surface -- the surface will be much rougher to visit.

Where will "hardcore" ambitious class tourists go? As in, what will be an exotic "off the beaten path" vacation for them?

Question: When does visiting the Moon get "ho-hum"?

Answer: After the Disneyland-style domed city is built.

To avoid that ho-humness, what can be done on the Moon that will feel exotic? What can be done outside the domed city?

For those that want an exotic someplace other than the Moon, what will be offered? They will want to go to Mars. Will they want to go beyond Mars?

Conclusion

I foresee that space exploring will continue, but beyond earth orbit it will continue to be a small scale hobbyist oriented activity. It will be small scale and robot-centric. The only activity likely to become commercial beyond earth orbit is space tourism.

Conversely, earth orbit activities are likely to continue to be commercially oriented and the activity in the earth orbiting realm is likely to continue growing steadily. As a result one of the growing challenges will be getting rid of the "space junk" -- obsolete satellites that continue to orbit.

Warfare

Introduction

"We are well prepared to fight the last war. The next one will be full of surprises." -- a military truism

Warfare-style conflict is going to steadily get faster and more decisive. The shooting part will become drones versus robots with very little human interaction. The battles will take minutes to days, no longer than that. And the wars they are part of will take days to weeks, no longer than that. Finding peace after the shooting, however, can take much longer, as it has in the Middle East in the 2010's.

The humans, the "boots on the ground", will be there to deal with other humans, mostly the civilians that get swept up in the conflicts. If this style of soldier carries a gun it will mostly be to uphold tradition. So, the shooting part of the conflict in any particular geographic area will be over in minutes, and then comes the succoring the civilians part, and that will last a lot longer and get a lot more news media time.

The winners of the robots fighting phase will be out in the open doing their succoring. The losers will be hiding and sneaking around. They will be looking for terrorist-style opportunities. This will be a time and place of pervasive surveillance, so the sneaking and terrorizing will have to be done in ways even more clever than those used today.

How will human performance-enhancing mix in? Even enhanced humans are going to be slow and clumsy compared to purely robot fighting machines, so there won't be Captain America-style warriors during the conflict phase. Can Captain Americas be more effective in the succoring phase? If so, that is where they will thrive.

More thoughts

Between the mix of wearables influencing patriotism and cyber making suggestions to people, how will patriotism, conflict and military play out? This is going to be very different from the 2010's. What will be the most the same is the joy of rallying around the flag. What will be the most different is picking up a gun and marching off to war. There will be a lot discouraging that from happening. An example being cyber saying "No! No!" and backing that up by not letting the Big Business they control finance the military adventure. If someone wants to go to war, they are going to have to find a way to get luxury money and dilettante activities to support it. The result: if there is one it will be small scale compared to what we think of as a war in the 2010's. It will be more like gang war or guerrilla war than military-on-military war.

Pervasive surveillance is going to affect war making. The more you can see the more you can shoot accurately at. If the conflict is between two advanced powers it will start with drone vs. drone over the conflict zone to see who controls the airspace. That phase will end quickly with a winner. That side can then move soldiers and ground drones more openly. The loser side will have to hide in various ways -- the simplest being just become part of the crowd, as we have in terrorism today. What other ways?

These conflicts I'm envisioning are not going to be total war affairs with nuclear mixed in. They are going to be like Middle East conflicts of today -- local regions with local issues, local governments, and lots of proxy-style intervention. The humans -- the boots on the ground -- will be there to succor other humans, not to be shooting at enemies.

Additional thought: the hiding side may use mosquito-size drones that can go undetected for a while. Maybe, but as I think about it, not likely because any control signals back and forth can be easily traced.

Electioneering in VR

Introduction

In the 2050's communications is going to be quite different from the 2010's. One change is that most people will have internal communicators – smartphones will become internalized. Another is that people will talk a lot with cyber muses in addition to other people. Yet another is that people are going to spend a lot of time in VR environments, and they will do a lot of communicating while in those.

One activity this will effect is campaigning for elections, and that is what this essay is about.

Hyper Image

What a person looks like to other people while in VR is going to be very arbitrary. It will have little to do with what they look like in the real world, and it can be designed. VR image design will become another dimension of the world of fashion.

Politicians will be taking advantage of this even more than most people do. They will be working hard at looking convincing to the constituency they are talking to in VR, and no surprise, that look can change with each constituency. It can also change with each bureaucrat, lobbyist, other politician, and influence group they talk with as well. Whew! Lots of change, and a place to spend lots of effort and attention on. The politician's image staff is going to be a busy group.

This means that a politician's staff's image-shaping energy is going to be split between two broad categories: researching what the voters want to hear their person say on the issues, and researching what visual image these voters want to see of their person in VR promotions and sound bites.

My Fellow Americans...

Beyond creating the images, a major activity will be getting these VR images in front of the VR images of the politician's constituency. They will do this to both convey their message on current issues they are making decisions on, and to get elected in the next round of election cycles.

This image can be presented directly by the politician, or they can have a specialized variant of a cyber muse present it. This specialized variant will appear to be the real politician. (in their VR form, of course) It will be programmed to make a presentation, then take and answer questions as if it were the politician. Since politicians like to appear consistent in their views on issues, this Q&A part of the programming should be very easily accomplished.

25 hours in a day

The biggest virtue of VR campaigning is that the politician-plus-muse can be in many places at the same time. The VR image can be in many places and taking questions from many people. The staff and their muses can be digesting these Q&A sessions to adjust the campaign rhetoric on a day-by-day basis, or even faster.

Robot Companions

Introduction

One of the upcoming high profile uses for lifelike looking robots mixed with artificial intelligence is as affectionate companions. These should become even more popular than pets because they can "outpet" pets in the sense of being even more faithful and affectionate and with lots less mess, maintenance and tragedy.

These are first generation, they will be followed by even smarter companions -- virtual and physical -- who will become what I call cyber muses -- robots specifically designed to inspire humans to achieve what their humans want to accomplish. The cyber muses will come about after AI becomes both self-aware and self-designing and moves well beyond what human designers have accomplished.

Companions

Humans like companions. But companions are a mixed bag: they can provide lots of support and they can provide lots of problems and issues. One current solution to the issues problem is to get a pet instead of a human companion -- pet relations and maintenance are a lot simpler. An upcoming solution is a robot companion. These will be even simpler than pets to maintain and the relations will be much more customizable.

These companions will have quite a range in design and relations with their humans. On one end of the range they will be quite simple in design and mostly conversational -- they will be virtual companions. On another end they can become quite physical and quite physically helpful as well as conversational -- they will become the "robot butlers" that have been a science fiction dream ever since robots were thought of. Yet another end will be "super pets" -- these designs will imitate cats or dogs only do the companionship part much better than cats or dogs do.

One of the high-profile early companion applications for human-looking robots mixed with some simple AI is being a sex toy. This is simple enough to do that a few have been designed in the 2010's. This is one of the earliest applications because it is so simple to do and the instinctive thinking supporting it is so strong. These get around what can be big frustration issues when dealing with real-world human companions.

Surprise Use

Another early application for human-looking and talking robots is being a Booth Babe at trade show exhibitions. This is a surprise application for me, LoL! But after thinking about it, it makes sense. The human Babes don't need to know their product in depth, they are there simply to attract people with their appearance and personality and then explain product basics so exhibitors get interested in finding out more. It is a monotonous job which makes it quite eligible for automation. Bonus: the android Booth Babes will know their products a whole lot better than the 2010's human Babes do.

Conclusion

Robot companions are a highly anticipated mix of robotics and artificial intelligence. They have been anticipated since the early days of science fiction writing. How they will appear in real life will be much different and much more diverse than those early envisionings -- they will first appear as conversational companions and pets and only later as physically helpful butlers.

What they will do well is be affectionate companions that can stroke human comfort emotions reliably without much mess or risk. The more they can do this effectively, reliably and cheaply the more popular they will become.

The Evolution of Wearables

Introduction

In the 2010's wearables are devices that you can put on your body that monitor bodily functions such as blood pressure, breathing and motion. They are of interest mostly to exercise and good health enthusiasts, but are also useful in a few health care applications.

This is just the beginning.

I envision the range of tasks wearables can be used for as growing enormously as technological improvements let wearables measure many more different bodily functions, and not much later, control them.

This change in what wearables can do between now and the 2050's is the topic of this essay.

Early Stage Capabilities

The early wearables will be bulky and concerned mostly with monitoring. The variety of what is monitored will grow steadily with time. Big breakthroughs will happen when parts of the monitoring system can be inserted into the body on a long-term basis. If they are just skin-deep this makes blood monitoring much easier and more versatile. If they can be lodged in organs that are deeper in the body then those organs can be monitored, both with more precision and more versatility.

The 5 Oct 17 WSJ article, The New Workout Secret: Clothing Sensors That Up Your Game by Lucy Danziger, describes how 2010's wearables, first generation wearables, are influencing exercising.

From the article, "I looked down at my iPhone's Nadi X app, which pairs via Bluetooth to the pants. The app, designed to teach you yoga poses or guide you through a flow, suggested that the gentle zapping meant I needed to ground my heels and push my tush back and up—something patient instructors have told me without resorting to tickling my ankles."

Later Stage Capabilities

The decisive later stage capability is adding adjusting internal conditions to simply monitoring them. As with the monitoring, this will start with a few simple items, then get more diverse and subtle as understanding and technology improve. An example of adjusting that is likely to be implemented in the early adjusting stages would be controlling blood sugar levels in diabetics.

What difference will wearables make?

We have talked about what wearables can do. Now let's look at how they will change our lifestyles.

Commodity uses

The commodity uses are going to be helping health and maintaining better awareness of the body's condition in general. With time what is monitored is going to get more varied and more nuanced. This will make keeping good health easier and easier. There will be fewer surprises, fewer scary crises.

It will also lead to smart nagging and shaming. This will evolve from the general-style of nagging and shaming that is already routinely engaged in by health insurance companies of the 2010's -- they send out lots of letters and robocalls about good health rituals a customer should be engaging in.

With wearables the nagging and shaming can get much more directed, specific and real time. An example: you are sitting at a restaurant with a hot date, and while you are gazing into those dreamy eyes across the table, you are getting a hint through your internal communication system, "If you eat that second helping of kale, you'll be three more pounds overweight and it will take two days of jogging to work it off."

Surprise uses

Recreational Mind Altering

One of the big surprise uses for advanced wearables is going to be mind altering. The more wearables can adjust hormones and the involuntary nervous system signals the more they can adjust emotions. If you see something scary, but don't get a surge of adrenalin because your wearables are suppressing it, will you feel scared?

Here is a more socially significant example: A person gets drunk because alcohol is affecting the brain. If a wearable can affect the brain in a similar way, it can get a person drunk too.

The big difference between wearables and chemicals is that ultimately -- when they get advanced -- wearables can change their effects a lot faster and be a lot more wide-ranging and subtle in their effects. If wearables are causing the drunkenness a person can sober up in minutes, not hours. And how they feel drunk can be adjusted -- it can have flavors.

As an example, a college student at a raucous frat party can be drinking water and getting plowed out their mind as the night goes on. Whoopee!! Then when party-is-over time comes they can sober up in about ten minutes... while they are walking back to the dorm or being driven home in the driverless car they called for... and get in an hour of homework for the next day's classes before they call it a night. Quite a difference in lifestyle.

And this is just a first-generation use of mind altering. Here are some other possibilities.

Reducing fear of flying

An example of something 2nd generation and subtler would be helping people overcome their fear of flying. When most of the people who have to fly are no longer scared of flying, commercial air transportation systems can get a lot more practical and a lot less ritualized. The first high-profile change would be transforming the TSA away from the neoreligious ritual it is in the 2010's into something much lower profile and much more effective. Another change would be letting airports get smaller and more dispersed so there is a lot less congestion. In sum, losing this fear would let a lot of airport and airplane redesign happen. The change would be dramatic. Ultimately, getting on a commercial airline flight could become as simple as getting on a bus is in the 2010's.

Reducing NIMBY

"No. Don't make that change to my neighborhood. It will kill my property value."

This is feeling NIMBY -- Not In My Backyard. It is a feeling that dramatically slows down changes in established residential areas. Related is the bad feeling that comes with forcing Grandma out of the apartment she has lived in for twenty years. The result is that once an urban neighborhood has been developed it can stay pretty much the same for as long as a century. This slows progress and creates "Rust Belt" cities. When NIMBY is strong neighborhoods get abandoned, not changed.

Part of what is happening here is fear, the fear that change in the neighborhood will be for the worse. If this fear can be diminished by wearables then lots more progress can come to established urban neighborhoods, and that will make a big and beneficial difference in how cities evolve.

Unleashing Nuclear Energy

Public opinion of embracing nuclear energy has been deeply and enduringly cursed by its first use: as an atomic bomb way back in 1945. The result has been widespread and deep fear of using nuclear power. A 2010's example of this enduring fear is the still-widespread concern about radiation coming from the Fukushima nuclear accident six years after it happened. If this instinctive fear can be diminished by wearables then nuclear energy can be used more widely -- widely in this context meaning widely as electrical power stations and widely in other applications. We may finally see lots of nuclear powered vehicles, and even artificial hearts.

In sum, as the wearables get into their third generation capabilities, all human activities that are partly emotion driven can be transformed -- gut feelings of all sorts can be controlled.

Quite a change.

Conclusion

As wearables develop and become widespread in society they are going to transform how people live in dramatic ways. Their commodity use is going to be improving health. Their first surprise use is going to be letting people indulge in high tech mind altering -- high tech getting drunk and high. Their second surprise use is going to be controlling emotions. They are going to let people control what they are fearful of, and what they love.

The world of wearables is going to be as different from 2010's living as Agricultural Age living is from Industrial Age living.

Lots of big changes are coming.

Further Reading

- The New Workout Secret: Clothing Sensors That Up Your Game by Lucy Danziger, Oct. 5, 2017 - https://www.wsj.com/articles/when-your-gym-shorts-tell-you-to-work-out-harder-1507215619
- Wikipedia: NIMBY/Not In My Back Yard - https://en.wikipedia.org/wiki/NIMBY
- Wikipedia: Fukushima Daiichi nuclear disaster - https://en.wikipedia.org/wiki/Fukushima_Daiichi_nuclear_disaster

Personal Performance Enhancement and Wearables

Introduction

I envision wearables as going through three generations of enhancements:

- The first generation will monitor basic and easy to measure functions and report them to the user. This is what we are experiencing in the 2010's.
- The second generation will monitor many more bodily functions, report them more quickly and widely, and begin to do some primitive controlling of some of the functions they are monitoring. A current example of controlling is a heart pacemaker.
- The third generation will monitor and control lots of subtle bodily functions. One will be hormone levels. This will let the wearables influence emotions such as fear and love. This ability opens wide possibilities in influencing how we think and live.

This essay will talk about second generation wearables. These are wearables that can monitor a lot and influence a few basic bodily functions. How will this level of control influence how we live?

Personal Performance Enhancement

Personal performance enhancement is going to be the core of why people get second generation non-medical wearables. Some of the abilities they will enhance are:

- better physical performance -- be stronger, better coordinated and have more endurance. This will lead to things such as being able to do sports and dancing activities better.
- better control of the sleep/awake/meditation cycles -- stay awake longer when you want to stay awake, go to sleep faster and reliably when you want to sleep, arrange nap times more easily. And when you sleep it is more likely to be refreshing, not restless. And likewise, when it comes time to meditate that state can be reached more quickly.
- enhanced senses -- all the senses can be enhanced, and perhaps new ones added.
- enhanced communication -- internalized communication and better assimilating what is communicated, as in, filtering out the nonsense and useless chatter.

The above are the obvious ways the 2nd generation wearables are going to enhance. Now some details.

Physical performance: People like to move around better. Helping to make this happen is why these wearables will be popular. The first stage is measuring, that has started now and will become more extensive as the measuring technology improves. This is popular because it strokes the "How am I doing?" emotions. It won't come as quickly but at some point the wearables will be able to modify as well as measure. Again, an example of this is a heart pacemaker. The modifiers will affect muscle growth, coordination and efficiency (which will show up as more endurance). These changes will be happening in tandem with better medical care wearables. As an example, walking better will be a goal of old-age medical wearables. When that is well accomplished the sports enthusiasts will be wondering if running and other sports performances can be improved by making simple modifications to the technology.

Attaining restful states: People love to work hard, and they then love to rest and relax well between the hard work. Wearables will help people get into these restful states more quickly and deeply and with more certainty. Refreshment will come much faster. One surprise consequence is this will change storytelling -- a character in a story can't signal that they have a problem by reporting they can't sleep well.

Enhanced senses: The enhanced senses will be things like first having clearer vision in the visible spectrum, then enhancement by seeing in UV and IR. Hearing will be hearing beyond the normal human range. Then we have touch, taste and smell.

How much of this will be handled internally, versus wearing VR goggles of some sorts, will change over time. Basically, the more exotic senses will start in goggles and as mastering them becomes more routine they will be internalized.

A new sense would be something like adding seismic sense, as in, hearing or feeling lower frequency vibrations than are normally felt. There is a limit to how low your ears can hear, but your skin can sense frequencies that are lower. These feel like vibrations. What can be changed is to have them feel more differentiated and more emotional, more like low frequency music.

Other conventional senses: Experiencing better taste, smell, touch.

A big question for these style of enhancements is what are the advantages, other than being a fussier dilettante at meals? What will these enhancements let people do that makes a difference?

Answering this is a big challenge. They are competing with AI and cyber to do things which make a difference. The more the answer is "not much" the more they are relegated to enhancing dilettante activities. An example of attractive dilettante activity is superior coordination which makes yoga, gymnastics and sports much easier.

Losing Performance

Here is a surprise: not only can wearables enhance they can also weaken, and this weakening is already happening with the first generation devices – smartphones.

The 6 Oct 17 WSJ article, How Smartphones Hijack Our Minds

Research suggests that as the brain grows dependent on phone technology, the intellect weakens by Nicholas Carr, talks about how current smartphones are reducing human performance on thinking tasks.

From the article, "Scientists have begun exploring that question—and what they're discovering is both fascinating and troubling. Not only do our phones shape our thoughts in deep and complicated ways, but the effects persist even when we aren't using the devices. As the brain grows dependent on the technology, the research suggests, the intellect weakens."

Driverless People

Something of a surprise: having "driverless people", as in, people engaging in enhanced multi-tasking during daily activities. These driverless people will be using their enhanced senses and doing lots of communication while engaging in day-to-day activities such as walking down the street. The wearables will be equipped with alarms alerting the person when they are "doing it wrong" in ways that wearables can detect. The wearables will help these people stay out of the way of cars and light poles as they do their walking down the street.

Conclusion

Wearables are first going to make a difference by doing better monitoring of a person's health. These are first generation wearables. The second generation wearables are going to do modifying as well as measuring. These will begin to affect physical performance and sensory awareness. They will be interesting to people because of these enhancing abilities. The third generation wearables will be able to monitor and control subtle things such as hormones and instinctive thinking. They will make an even bigger difference, but will take a lot longer to develop.

Wearables are going to make a difference, and that difference will be constantly changing as wearable capabilities constantly change. This means people will be constantly learning right ways and wrong ways of using the wearables.

Further Reading

- How Smartphones Hijack Our Minds by Nicholas Carr, Oct. 6, 2017 - https://www.wsj.com/articles/how-smartphones-hijack-our-minds-1507307811

Mixing Romance and Wearables

Preliminary Thoughts

First, keep in mind that romance and marriage are going to be changed dramatically from what they are in the 2010's by the Tattoos and T-Shirts issues. Their roles in society are already changing and will continue to change. Lets look at each individually.

Marriage

Historically, the commodity use for marriage was two-fold: 1) to produce children who would get support as they grew into adults, 2) to establish kinship linkages between extended families so they would trust each other more and could cooperate with each other more. By the 2010's the second use had become irrelevant because cooperation is being handled in so many ways outside of marriage connections. This is why arranged marriage is anachronistic in the US. And even marriage arranged by the couples themselves is losing some relevance in regards to child raising. Thanks to increasingly widespread prosperity and government supported social support networks the child raising use is also not as compelling as it used to be, either -- we now have lots of single mom's.

This evolution into anachronism is going to continue. By the 2050's both of these activities are going to be handled by other groups and organizations. This means that what is left for marriage is going to be manifestations of personal expression. It will be sustained by the same style of thinking which today sustains horse riding, oil painting and violin playing.

This means:

- *It is going to be done more perfectly by the participants.* People are not going to be in a marriage to satisfy the wishes of other people. And because divorce is easy they will also not be in one if they think they have made a mistake. They will be doing it right in their own eyes, and doing it right, right now.
- *There is going to be more ritual and less practical.* Personal expression supports a lot of ritual. Being married is going to be a lot more elaborate than just being together.
- *Participating is going to be a lot more expensive, so fewer people will be engaging in it.* Weddings and marriages are both going to get more elaborate and more expensive because of the desire to support personal expression.

An interesting related phenomenon is that in the 2010's some women are going through marriage rituals without getting married to anyone. (mostly in Japan and Korea at this point) These women just really want to experience the ritual.

In sum, the reasons for engaging in marriage are going to change dramatically. Marriages will be based a lot more on personal expression and a lot less on supporting child raising or uniting families.

Romance

Romance is going to change as well. The commodity use for romance was spending time with someone to see if you wanted to spend even more time with them, with having sex or becoming married being a large part of the time together theme.

As marriage has changed its role, and as birth control technologies have improved, romance has changed its role as well. First, let's look at romance's purpose, which is to answer a couple big questions:

- What is a person going to be looking for in a partner?

- What will they do with the partner?

The commodity change that has happened over the last fifty years is that romance is a lot less risky to engage in. As marriage's role has changed and become more "perfect" and more about personal expression, romance's role is changing as well.

Here is how the shift towards personal expression is changing romance:

- *In some social situations the routine is simplifying.* The "hook-up"/"Netflix and chill" culture is an example of this happening.
- *The ritual that hasn't gone away is the social shaming when marriage and romance with someone outside the marriage are mixed.* In high profile cases, such as moralizing politicians and religious leaders, it has gotten even stronger.
- *What has gotten more expensive and more elaborate about romance?* The more romance becomes about personal expression -- the trend that is also affecting marriage -- the more elaborate and expensive it is getting. And the more ritual gets involved.

As romance becomes more about personal expression and becomes more ritualized, being "just friends" is going to take over activities that are considered romantic in the 2010's. In the 2050's being just friends is going to include a lot of intimacy. There will be a difference between being "friends" and "just friends" and there will be some language changes to help define the differences.

Here are some places where being friends is going to take some more word defining:

- *How many friends will a person have?* For instance, just one partner? Or will many partners become commonplace and more acceptable?
- *Related: how are all the "stepping out" forms going to be viewed?* What forms will infidelity take? The forms should be quite different. And the opinions of those around and watching will be quite different.
- *What will be the root satisfaction, the form of personal expression that is giving deep satisfaction?* To help answer this, think of what satisfaction people in the 2010's are getting from tattoos?

These are some of the basic romance issues. Now... let's mix in technology. Let's mix in... wearables!

Mixing in wearables

Wearables are going to make romance quite different in the 2050's.

Background

In the 2050's we will have both emotion-controlling wearables and cyber muse companions who make lifestyle suggestions to their owners.

The emotion-controlling wearables mean that there is no uncertainty about "Will I like this person?" now or in the future. If you want to, yes, you will. The wearables will see to that. Conversely, if you don't want to fall for someone, the wearables can help you "forget him" quickly and decisively.

The cyber muses are interested in getting people into a parental state of mind -- any child that is not conceived by and raised by humans in some style of family structure (including baby clubs) will have to be created as a Necessity Child (in a vat somewhere) and raised by cyber nursemaids. There will be plenty of these being created in the 2050's, but the less needed the better.

But the muses are usually subtle about this become-a-parent suggesting -- people don't keep cyber muses around to convince them to become parents. They are there because "Behind every great person there's a good cyber muse." People want to be inspired, not nagged.

The big change between the 2010's and the 2050's is that deciding who to fall in love with is not a question of emotional feeling -- because of wearables that attraction will predictably follow after the choice is made, and the emotion will be deep and lasting, and guaranteed as long as the person uses wearables.

The "engagement ring" of this era is programming liking a person into your wearables -- or giving your object of affection a new wearable that will handle the task of making them like you. And, as mentioned above, not falling for someone is equally easy to arrange.

Given this environment... what *will* be the hard choices in "getting committed"? What will people think hard about? What will they look for, long and hard, in potential partners? What kinds of socializing will they do to size up potential partners? This new style of socializing will be what people think long and hard about.

It will be quite different.

Practicing romance with wearables

Using wearables for romance is something teenagers will practice with, just like they practice with alcohol, drugs, music, sports and driving. They can practice controlling emotions and other mind altering with wearables.

How to practice with them will be a hot, opinionated topic of the day. It will be much like how to practice with drinking and dating is in the 2010's.

Part of what makes it so white hot is how fast the emotion controlling capabilities of wearables will change as the technology advances -- parents' wearables are not going to operate the same as their kids' wearables do.

Conclusion

Marriage is going to continue its evolution away from being a social institution for child raising and unifying families. It is going to become more and more an institution that allows personal expression. People are going to get married so they can have a perfect marriage with a partner who also wants a perfect marriage. This perfection won't require children or be about creating the right in-laws, but it is likely to be about engaging in expensive rituals that both partners enjoy.

Likewise, romance is going to change as well. Because of wearables it will no longer be about answering the question, "Can I fall in love with this person?" -- wearables can decisively answer that question simply by dialing in the answer the person wants to have. Romance is going to be about answering the question, "Do I want to fall in love with this person? I know I can if I decide I want to." It is going to hark back to arranged marriage thinking, but with the person asking the question being the one doing the arranging.

Further Reading

- Tattoos and T-shirts: a study in how High Tech replaces Low Tech by Roger Bourke White Jr., August 2002 - http://www.whiteworld.com/technoland/stories-nonfic/2008-stories/Tatoos-Tshirts.htm

Desire and Wearables

Introduction

One of the choices wearables will offer is not feeling sexy at all. People can choose to adjust both: not being interested in the sex of other people, and not being interested in their own sexuality.

It has happened before

The historical antecedent for this is eunuchs. These are sexless males and they date back to the dawn of history. The big difference I see in the 2050's version is it is reversible (dial up and down this wearable choice) and it can happen in both sexes.

Many historical eunuchs suffered this fate as a form of shaming. They were castrated, enslaved, and forced into hard labor projects. A few became servants to high level royalty. A few of these high level servant types were chosen for their role in boyhood -- their role was to insulate the royalty from scheming servants and other people who had political or dynastic agendas. Also chosen in boyhood were choir singers who would then gain unnatural but interesting voices.

Another form of celibacy, a voluntary form, is being a cloistered priest or monk.

The 2050's version

None of the above will be the role for 2050's wearable-induced celibate people. They will be sexless because they personally see lots of frustration with feeling sexual. Think of the frustration expressed in songs and stories about unrequited love.

How will this choice impact how a person lives?

The obvious first use is to avoid distraction and frustration. People can concentrate on what they consider the important issues in their lives. This will make them like the priests and monks who choose to live in monasteries and temples to avoid worldly distractions. But in the 2050's world a physical retreat will not be necessary -- dialing up a wearable choice can be sufficient.

What other lifestyle choices will be affected?

Wearable-induced celibacy could become a treatment for sex problems and a punishment for sex offenders.

Fashion choices will feel the effects of this. People dress for themselves and other people. If they aren't interested in looking attractive in a sexy way this will make big differences in what seems to be attractive clothing, and how often clothing needs to be updated to stay fashionable.

Socializing will change too. Singles bars will lose some patrons or add other themes. Raucous entertainment may diminish in attractiveness -- I'm thinking of frat parties and girls enjoying fainting at popular music concerts.

What are other surprise consequences? Example: How much does wanting to raise a child depend on sexuality? Will Wearable-induced celibate people want to do this just as much as sexuals? (palace eunuchs raised royal children) If they do, what will be the differences in their styles of child raising? Will they raise celibate children who look upon sexuality as a distraction? Will they raise hypersexual children because they can't easily distinguish normal behavior from hypersexuality?

Conclusion

Wearable-induced celibacy is likely to become a common lifestyle choice in the 2050's. The lifestyles it supports will be quite different from the historical eunuch and cloistered monk lifestyles that supported celibacy in the past. How it will come about, and how it is practiced will be surprising compared to these historical antecedents.

Us versus Them Thinking and Wearables

Introduction

Us versus Them thinking is powerful instinctive thinking among humans, and it influences how our communities develop in many areas. Things such as neighborhood composition, land use, employment, racism, religion and politics are all heavily influenced by this instinct.

There is a lot of both love and fear mixed into the Us versus Them feelings. Patriotism is a love feeling. Worrying about strangers stealing women and chickens is a fear feeling. Politics goes both ways -- Us versus Them emotions swirling is why many strange laws get proposed, and some pass. Sports events are about loving one team, while being a hooligan is about hating another. Becoming an employee is much like a romance -- will this person become one of Us, a good one of Us? When things go wrong the common emotion is Blame Them. And there are all the emotions swirling around religion.

If wearables can be used to influence how this Us versus Them instinct manifests itself then humanity has gained a powerful tool in the battle against material waste and making silly, expensive choices in the social sphere.

Instinct origin

When mankind lives in Stone Age conditions he lives as a small collection of people, a proto-village. The group typically consists of one-to-a-few extended families. These people live and work together, and when moving time comes they move together -- Stone Age living is a semi-nomadic lifestyle.

This means that these village people are treated differently than the other collections of people that are living around them in the region. These people are family and those other people are strangers. Mankind has been living this Stone Age lifestyle for tens of thousands of generations, basically from the beginning of mankind until the invention of the Agricultural Age which began in a handful of river deltas about five thousand years ago, an eye blink in the evolutionary timeframe. In Agricultural Age lifestyles a larger collection of people gather together and living conditions become very different. This means that evolution has had plenty of time to hardwire human thinking to support Us versus Them thinking, and not much time to lose it in those situations where it is no longer appropriate.

In sum, it is a well established instinct. And it is around whether or not it makes sense for the living style a person is enjoying.

Trusting and betraying

The heart of this instinct is deciding who to trust and who to betray. If the person is family, you trust them, if the person is a stranger it is OK to betray them -- betraying doesn't create much guilt feeling, and it can create great boasting stories. As a person one meets moves from feeling like a stranger to feeling like family then betrayal becomes less comfortable. This is what initiation rites and getting to know someone over a meal are all about. They are moving the person from stranger to family in the eyes of the Us versus Them instinct.

Conversely, social shaming and bad gossip are about moving the target person out of the family side of the instinct and into the stranger side, so that betrayal becomes comfortable.

What wearables can do

Since Us versus Them is instinctive thinking, emotional thinking, it may be that wearables can be designed to influence it. As an example, if a person starts thinking about betraying someone else -- say, robbing a store -- the wearable can stimulate powerful guilt thinking and the person doesn't think it is such a good plan anymore.

Business transactions bring up a lot of Us versus Them thinking. Here is a classic situation: A person at a negotiating table is thinking, "Do I think this deal is fair? If I don't, and I think I'm getting too much, do I confess and make changes to make the deal fairer? Or do I just take all I can get and laugh all the way to the bank?" Dealing with the emotions of these kinds of situations is what makes deal-making a distinctive form of thinking that some people are much more comfortable with than others. And this kind of situation is muddy enough at the emotional level that it is unclear what a wearable would add to this kind of deal making process.

Which brings up the next point.

It won't be easy

Compared to romantic love or fear of flying Us versus Them thinking is going to be much harder to detect, and the object of the thoughts is going to be much harder to determine. Controlling Us versus Them is going to be a late stage development in the wearables-controlling-emotions spectrum.

But it will be an important one because cooperation with strangers is so widespread and so important in societies based on advanced technologies.

Conclusion

Us versus Them thinking is a powerful instinct. It is one that is at the root of deciding whether to trust or betray when dealing with another person or organization. It is an instinct that is well-suited for Stone Age lifestyles, but doesn't work nearly as well in more technologically advanced lifestyles. In these lifestyles cooperation with wider and wider groups becomes more advantageous and betrayal become more damaging.

Where wearables can help in reducing damaging betrayals they are being of big help in the communities of the 2050's.

Updating hardware and software

Introduction

We are experiencing an ongoing crisis in updating hardware and software: How to keep the hardware and software behind all this Big Processing and Big Data up to date?

We have this crisis already in the 2010's and it will still be with us in the 2050's: Hardware and software change so quickly that 20 year old equipment and programs feel ancient. But in many instances they are still used. A high-profile 2010's example of this happening is in US air traffic control and airline scheduling systems.

This is going to stay a problem, but solving it is something that can be included in structure design, if it is thought about ahead of time. Basically, it means adding redundancy to the design so that parts of the system can be shut down and outright removed without disrupting the whole operation. Redundancy is expensive, it adds cost, essentially doubles it, but it is another form of insurance. This means it looks expensive until the time comes when it is helping to save the system from catastrophic and anachronistic circumstances.

How self-aware AI systems -- post-"Singularity" systems -- will handle this will be interesting. It will be particularly interesting if these AI systems develop complacent thinking, as in, "I'm thinking just fine! Why do *I* need an upgrade?" Keep in mind that throughout history complacent thinking has been the enemy of progress and friend of decline. There is no reason to believe this trend will end in the self-aware AI realm.

Social Revolutions

This second section of the book is about the social revolutions that the new technologies of the 2050's will support. This is about how people will change their activities in response all the neat new gadgets and technologies that will be available to them.

Three Kinds of Money

Introduction

One of the big changes the 2050's will bring is how money is acquired and spent. First off, this will become a cashless world -- transactions will be conducted using biometrics and electronic accounts not paper, plastic or coin. This is just the beginning of the changes.

I envision three kinds of money emerging: Necessity Money, Luxury Money and Investing Money. These forms will be used for purchasing different styles of goods and services and will be difficult to interchange. (if they are easy to interchange, then there is de facto only one kind of money)

First off... Why have money?

The root purpose of money is to make it easier for people to cooperate with each other. When people can cooperate more easily all sorts of benefits accrue to the people cooperating and to the communities around them. Money is all about cooperating. This is why we have money and why its forms change over space and time.

The earliest form of money is oath trading. "I promise I'll do this for you if you promise you'll do that for me." It is simple, but the range of tasks and conditions a promise can handle is limited and the details of what is being promised can be confusing and become a source of disagreement. Money is a way of getting beyond oaths' limitations. It supports many more styles of cooperation.

Who gets involved?

Because money is all about facilitating cooperation lots of people get involved with it. Money is a tangible form of cooperation. It is easy to measure and easy to trade. Another benefit is that it allows cooperation from the past and future to be traded. A person's savings are past cooperation, and mortgages are about future cooperation.

Advancing cooperation

As money gets more sophisticated, trading gets easier, and new kinds of cooperation can be supported. In the case of house building, add in a mortgage and the cooperation gets more convenient and what can be constructed gets more complex. Investing in a company is much easier when there is a stock market. This means that the variety of companies that can be successfully created grows as well. A successful steel company requires a lot more cooperation than a successful Main Street grocer.

And betrayal

We have money so we can cooperate, but it also supports betrayal. And, like cooperation, the more kinds of money we have the more forms of betrayal get supported. Oath breaking is the oldest style. "Your money or your life!" gets supported when tangible money gets common. Ponzi schemes are an example of a 20th century betrayal style.

Emotions are involved too. Being greedy is seen as betrayal in the eyes of those seeing the greed. It is seen as duly earned reward in the eyes of those who don't see the greed.

Just as there are many ways to cooperate there are many ways to betray.

More coming: Blockchain

Creating new styles of money is still going on. Paying with smartphones is one example and another is blockchain technology, of which Bitcoin is an early example. It is going to be interesting to see how these change how we cooperate over the next couple decades.

In sum...

Money makes the world go round by making many more forms of cooperation practical. And the more forms of money we have the more forms of cooperation are supported. As new kinds of monies are developed the world goes round in a much flashier fashion.

In the 2050's we will have even more new styles of money and that is what gets covered next.

Necessity Money

Necessity money will be the medium of exchange used to acquire goods and services offered by the Total Entitlement State (TES) system. TES will be providing basic food, housing, healthcare and dignity to everyone in the community. Everyone in the community gets an allotment of necessity money, but this doesn't mean everyone has to be purchasing the same stuff or in the same quantities -- people will still be able to make choices. Necessity money will be an evolution from the 2010's SNAP, food stamp and welfare systems. The money will be handed out on a regular basis.

There will be lots of monitoring going on in this necessity money environment. Many of the community will be paycheck-to-paycheck lifestylers and the system will be watching to make sure what they are purchasing is not going to cause a run-out-of-money crisis. There will also be lots of suggestions coming from the monitoring system. Ideally, the suggestions will decline as the person accumulates a surplus of unspent money -- their "rainy day" fund is large.

Want more? Wait for it

One way of balancing between the desire for goods and the money allocated to a person is to have the person wait longer to get things when their money supply is low. A person goes into a restaurant and orders lunch. If they have lots of money in their necessity money account the meal comes out quickly. If they are scraping the bottom of their barrel the food will take two... five... ten minutes to arrive depending on how close to the bottom they are. This won't be as inconvenient as this sounds because waiting is something that will be part of the TES necessity lifestyle environment. Think of waiting in line for rides at Disneyland. Instead of an inconvenience this will become a signaling device for many people to indicate the status of their money supply.

Luxury Money

Luxury money will originate from doing ambitious class activities -- the 2050's version of "real job" activities. It can be used to purchase anything people or cyber choose to make -- anything that is not purchasable with necessity money, that is. The two systems will be distinct and independent. This will be so to keep the goods and services provided in the necessity environment stable.

Luxury money can buy dilettante-created goods and services. And this will be one of the pillars for engaging in dilettante activities. If someone wants gourmet style vegetables grown by a dilettante farmer, they will pay for them with luxury money. If they want to vacation in an exotic locale they will pay with luxury money. Conversely, if they want to go to a Disneyland equivalent (entertainment for everyone) that will take necessity money.

Investing Money

Investing money buys factories. Since cyber is doing the factory designing, building and buying in the 2050's, the money that does this will flow through cyber hands -- humans will rarely have contact with it. It will exist because it is a convenient way of handling prioritizing the projects that cyber will be engaging in to provide prosperity to humans.

Humans won't see it, but it will have a strong influence on the prosperity that humans experience.

The 21 May 17 WSJ article, The Quants Run Wall Street Now *For decades, investors imagined a time when data-driven traders would dominate financial markets. That day has arrived.* by Gregory Zuckerman and Bradley Hope, describes another step being taken in this evolution.

From the article, "Up and down Wall Street, algorithmic-driven trading and the quants who use sophisticated statistical models to find attractive trades are taking over the investment world.

On many trading floors, quants are gaining respect, clout and money as investment firms scramble to hire mathematicians and scientists. Traditional trading strategies, such as sifting through balance sheets and talking to companies' customers, are falling down the pecking order."

How Will People Save Money?

Saving is part of human instinctive thinking. It is a recent part but, like exploring, it is something that some people in the community will want to engage in. It will make them feel good.

How to stroke this feeling is going to take some inventing. If you don't have physical money, how can you put it under a mattress? What will having a "fat bank account" mean in the 2050's? For necessity money this will be growing the rainy day fund, but there is a limit to what makes sense to have in this fund. Beyond that limit, growing it is just stroking some instinct in the person that is doing it. For this reason large savings accumulations will probably be a luxury money activity, but what it will be beyond that isn't clear.

Gambling and Investing

Gambling and investing are powered by similar emotions. Gambling will be a lot more popular, but both are going to be activities that people will want to engage in.

Gambling games are simple to organize. Casinos are likely to be around in the 2050's. The challenge is coming up with what can be gambled? A person who gambles money is going to run out of it at some point. One answer is providing gambling money as part of the necessity budget. A person can gamble until that runs out, then they must stop until the next paycheck comes along.

Another answer, one that exists in the 2010's, is computer gaming. Playing games is a form of gambling that doesn't require either necessity or luxury money.

Investing is not going to be as simple to organize. This will be done with luxury money, but the details of how to do this in the 2050's are not easy to forecast. It will likely center on dilettante activities, but how it will be organized will be different than 2010's investing is.

Additionally

The 18 Jun 17 WSJ Editorial, Zuckerberg's Opiate for the Masses *If we get 'universal basic income,' the millennials will never leave our basements.* by Andy Kessler, talks about Mark Zuckerberg's vision of money and people's activities in the future. It is much like mine.

From the article, "At Harvard's commencement last month, dropout Mark Zuckerberg told eager graduates to create a new social contract for their generation: "We should have a society that measures progress not just by economic metrics like GDP, but by how many of us have a role we find meaningful." He then said to applause: "We should explore ideas like universal basic income to give everyone a cushion to try new things." Who wouldn't like three grand a month?"

This is describing what I call Necessity Money and Top Forty Jobs (dilettante jobs).

Conclusion

Money is something that will be transforming dramatically over the next few decades. It will divide into three general categories and each of those categories will be used to pay for activities that are distinct in the lifestyles of those living in the 2050's. Necessity money will buy the goods and services provided by TES. Luxury money will buy goods and services that are offered outside of the TES system. Investment money will be handled mostly by cyber and will buy the tools that cyber use to produce goods and services for humans.

All-in-all, handling money is going to be quite different than what we experience in the 2010's.

Further Reading

- The Quants Run Wall Street Now by Gregory Zuckerman and Bradley Hope, May 21, 2017 - https://www.wsj.com/articles/the-quants-run-wall-street-now-1495389108
- Zuckerberg's Opiate for the Masses by Andy Kessler, June 18, 2017 - https://www.wsj.com/articles/zuckerbergs-opiate-for-the-masses-1497821885

Education: Necessity Style and Ambitious Style

Introduction

In the 2050's education is going to divide sharply into two general categories: Education for those who will be leading necessity lifestyles and education for those who will be leading ambitious lifestyles. What students will be experiencing in these two formats will be quite different.

Necessity Education

Most of what necessity students will be learning in their schooling, entertainment and recreation is urban legend. Why not? It is comfortable and it won't affect their well-being very much because cyber is controlling the material, service and transportation functions.

What will this lead to?

It can lead to education being used primarily for learning how to master the tools used for emotion manipulating. This will take the place of the 2010's goal of learning to master tools that will help in employment at jobs that make our world a physically more prosperous place. (Again, cyber will be handling that in the 2050's, and they will be assisted by ambitious class people who are learning very different skills than necessity class people are.)

If manipulating emotion is the core of what necessity style education teaches then what kinds of emotion manipulating will be valuable to learn? The two most influential categories will be praising and shaming.

Praising is learning how to pump up your buddies so they try even harder and do even better at those activities they care about -- these will be mostly dilettante activities such as sports and entertainment.

Shaming is about social censure. There will be lots of social shaming in necessity lifestyle environments. Shaming is likely to be a major time and attention consumer in necessity circles -- gossip will experience a strong revival.

Praising and shaming will be the core activities that necessity education will be teaching, and surrounding these will be lots of comfortable urban legend. This urban legend part will show up as the tendency for human teachers to editorialize on their subject matter. This tendency is going to be even stronger than it is in the 2010's. It will run rampant in the soft disciplines such as history and psychology, and it will also be strong in the hard sciences because people have hopes and dreams in those disciplines as well. A recent example I encountered of this wishing and hoping in hard science was several news articles in which quantum entanglement was being described as teleportation. Sadly for the wishers, dreamers, and the rest of us, they are not the same. The source of the excitement was a rocket ship launched into space that was demonstrating quantum entanglement with a lab on Earth.

Ambitious Education

The counter to this urban legend trend will show up in the education of the ambitious class. These are people who are being educated to make a difference in the real world, so it is really helpful if they are taught how the real world really works.

This aspiration to know how the real world really works will favor cyber-oriented education. The cyber teachers will find it easier to stay analytic and dispassionate about what is learned and how it is taught. The challenge is going to be finding humans who want to pay attention to this style of education. Again, these students will be ambitious young people, but they must feel that their ambition is being well served by paying attention to these teachers who will seem boring to most students.

One device that will help students stay interested in being ambitious is wearables. The "more interest in learning boring stuff" is an emotion the third generation wearables will be able to help keep strong.

Is interest in doing homework an emotion? I recall as a child being very fussy about what I wanted to spend time on learning, and moaning and groaning when I had to spend time on the boring stuff. The more learning is an emotion the more wearables should be able to influence it. The more it is an emotion the more wearables can become a substitute for learning the skill of self-discipline in study endeavors. It is possible that in the future we will be able to dial up an interest in finishing homework. That will help those who want to be ambitious and learn about how the real world works.

Conclusion

If learning urban legend satisfies the educational needs for most people it is going to spread widely because it is so comfortable. Most human teachers will teach the urban legends they are passionate about. They can do this because prosperity will be largely divorced from what humans learn.

If a person wants to be ambitious and shake up their real world, they will do most of their learning from cyber teachers. These cyber teachers will be the ones teaching rational, analytic-oriented learning.

How much wearables will be able to assist will be an interesting question. The more learning "boring" analytic learning processes is emotion-based, the more wearables will be able to help by modifying the emotions a person feels.

Health Care

Introduction

Health care is undergoing lots of changes in the 2010's. The changing is going to continue and likely get even more dramatic. By the 2050's things are going to be very different.

What will stay the same is that lots of ritual will be mixed in with the science and technology solutions. The rituals are there because they make people comfortable and sleep better at night.

This essay is about how technology and ritual will be mixing in the 2050's.

Adding technology to ritual

Health care dates back to prehistoric times. Helping fix a broken bone, helping a person eat and sleep while suffering from a recoverable disease, helping a young mother give birth, all these benefited the tribe living in Stone Age times. The tribe survived better. These Stone Age conditions lasted for thousands of generations, and these caring successes changed human instinctive thinking to embrace these activities. This is the source of the emotion supporting health care rituals.

Fast forward to the Industrial Age and the emergence of science, technology and scientific thinking. These started mixing with health care -- plaster casts, aspirin, and forceps helping a child down the birth canal are early examples.

As health care science and technology improved two big questions emerged:

- What really works and what is pseudoscience quackery?
- How to mix these new science-based methods with traditional rituals?

These are on-going questions that we still face in big ways in the 2010's.

Envisioning the 2050's

Some of the high-profile differences between medical technology in the 2010's and 2050's are going to be: artificial intelligence, pervasive surveillance and wearables that can modify body chemistry as well as monitor it.

But while there is going to be lots more science and technology in the health care mix, the rituals aren't going to go away -- they are still going to have a big effect on how healthcare is delivered.

There will still be instinctive demand for a lot of ritual -- "demand" in the "consumer demand for a product" sense. So, even if the combination of wearables and pervasive surveillance can bring "fast fixy" to people, they will still want rituals of many kinds entwined in the health process, such as visiting doctors for health advice. The rituals will be nutty when looked at from a cool-headed analytical perspective, but many health care participants will want them and think they are both very important and comforting. The rituals will help them sleep better at night... even though wearables will be doing that for them even better.

Ambitious and necessity health care

The ambitious and necessity communities are going to have different health care environments. Both are going to get the basics, and the basics are going to be really good compared to what is available at all levels in the 2010's. In this basic care arena much will be the same for both groups.

The biggest difference in overall health care is that ambitious people will have luxury money to devote to adding to what they get in the way of health care. Most of this difference will be devoted to cosmetic and performance effects -- luxury folks are going to look better and move through life better than necessity folks.

Another difference is risk-taking: How much risk can a patient accept when undergoing a treatment? Ambitious folk will be able to accept more risk.

Still lots of ritual

There is still going to be lots and lots of ritual mixed into health care. These days this tendency to ritualize shows up both in how patients and health care people treat each other and in how health care is paid for. The government regulations and health insurance practices are both just nuts -- both are held deeply hostage to The Curse of Being Important.

Ritual shows up in many other areas of providing health care -- think of the routines when visiting your doctor's office. The hard question to answer is: How much should these rituals be tolerated? Especially those that are expensive, exclude using effective technology, and are prescribed by onlookers, not the patients themselves?

Editorial: Embracing "Patient Pays"

How we pay for health care these days is just nutty. It is nutty in so many ways, and the alternatives that are emotionally embraced by lots of people in the community are even nuttier.

One alternative that I embrace is much simpler and more rational. I call it the "patient pays" system. Basically, people are given health care vouchers and they submit those to pay for treatments that *they* (the patient) choose to have. The patients choose, not the insurance companies or the government. This system would be lots less expensive (because of improved competition) and either more rational or better customized to what makes the patient feel good. This will be the patient's choice: If a patient feels better with a witch doctor performing rituals, why not? There is plenty of cheap technology also available to take care of the science-oriented issues that are causing the ailment.

In this system the important issue, in terms of choices made, is what makes the patient feel good.

Conclusion

Health care in the 2050's is going to have a lot more technology mixed in. These will include better health care devices, artificial intelligence and pervasive surveillance monitoring. The result is going to be a health care environment that is much more effective than that offered in even the best 2010's environments.

There will be differences in what people are provided. Those who have luxury money to devote to health care will be able to get things beyond the basics.

There is still going to be lots of ritual mixed in. These rituals are going to look silly to some people, but they are going to make others feel more comfortable and able to sleep better at night. This is why they will persist.

All-in-all, it is going to be quite a different health care world.

Further Reading

- The Curse of Being Important by Roger Bourke White Jr., October 2009 - http://www.whiteworld.com/cyreenikland/editorials/editorials2009/curse-of-important.html

Coming of Age Rituals

Introduction

What are coming of age rituals going to be in this era of driverless cars, cyber muses, TES, and wearables?

Driverless cars mean we will lose getting a driver's license as a widespread coming of age ritual. Likewise, those people who are living necessity lifestyles, not ambitious lifestyles, are not going to have getting a hard-to-acquire diploma and job as coming of age rituals. What will be the replacements for these?

Some Possibilities

Here are some possibilities for coming of age rituals for the 2050's.

Getting awards -- This has been a common ritual for many generations. The problem is: how much weight does the community give to the award? This varies all over the map from "[yawn] That's nice..." to "Wow! You have that! Let me open this door for you!" (as in, an opportunity) So if it is going to become a coming of age ritual, what needs to be clearly and widely established is how significant the award is. It must decisively rise above the crowd.

Peer pressure -- How is "I double dare ya!" going to be handled in an environment with both helicopter parents and pervasive surveillance? Related: How are gangs and gang initiations going to be handled? Will becoming a gang member still be an informal coming of age ritual? A possible new wrinkle: Will there be ways to be a VR/AR gang member, so you can join a gang while you're still at your desk in your parent's basement?

Military service -- Even though cyber and robots will be doing most of the actual fighting there will still be human soldiers. They will be mostly succoring the human casualties of the robot vs. robot fighting, but they will still be considered soldiers and they will still go through lots of training and rituals that are comparable to contemporary military training. Those who go through a stint of military service will receive recognition in their community. They will have come of age.

Religious service -- Gaining attainments in a religious order is another traditional form of coming of age that is likely to continue into the 2050's. Religion changes with time, geography and social circumstances so the attainments of the 2050's will be a bit different from those of the 2010's.

Having a child -- this can be a coming of age ritual that endures -- especially teenage pregnancy. Lots of emotion behind this one. And, probably, lots of cyber muse support. If the cyber muses are dedicated to keeping the human population up, and most people are urban dwellers with low fertility rates, they will be engaging in child-encouraging activities. It will endure but how it is conducted will be quite different. This will be a time when baby clubs are the social centers for most child raising.

Cyber Muses and Wearables -- How will cyber muses and wearables affect what are coming of age rituals? Will there be "child" and "adult" cyber muses? Will there be "child" and "adult" wearables? Will getting some adult versions take passing some kind of moderately difficult test? Will the difference be visible to others in the community?

Overall, what will kids have to work hard at, and accomplish, and aspire to accomplish, to be seen as having become adults? These will be the coming of age rituals for the 2050's.

Q: Will a coming of age event even be important in necessity social circles?

A: It will be important because this is supported by instinctive thinking. Because of instinct it will likely become even more important than in the 2010's. But, because of instinct, the rituals could look quite silly when seen in the context of harsh reality. Think of fraternity and college marching band initiations.

Coming of Age interacting with Tender Snowflake child raising

One of the big influences on Tender Snowflake coming of age rituals will be wearables. Wearables are important because they will be able to modify the fear emotions. How will Tender Snowflake child raising and fear modifying wearables interact? The Tender Snowflakes are going to be easy to scare, and their thin-skin outrage emotion easy to provoke. How much will wearables modify these? How much will they suppress? If they do suppress a lot of them, how will Tender Snowflakes differ from people who have learned fearlessness, tolerance and patience by education and widespread cultural experiences, and don't feel the fear or outrage, wearables or not? The two upbringings won't be the same, and how they will differ will take some thinking about.

And what will be coming-of-age rituals for Tender Snowflakes? What will be something they routinely engage in that requires a lot of sustained effort and uncertainty -- like getting a driver's license did in the 2nd half of the 20th century?

Conclusion

Coming of Age is going to change a lot by the 2050's. The world will be so different that coming of age rituals must also become different.

I foresee that communities all over the world are going to be searching long and hard for good answers to this question. And, it is likely the answers will be community-oriented -- as in, different for each community.

Baby Clubs

Introduction

In the 2050's baby clubs are going to displace marriages as the mainstream organizations for child raising. Marriages are going to evolve into being more about romance and couples expressing their love of doing many kinds of things together. Some couples will take on child raising, but most of the child raising will happen in a new style of organization designed specifically for child raising: baby clubs.

This essay is about what baby clubs will be like.

Purpose

Just as we have a style of organization that specializes in educating children -- a school -- in the 2050's we will have a style of organization that specializes in creating children and raising them -- a baby club.

Baby clubs will evolve as single parents learn to gather into groups and pool their resources. They will come to be encouraged by communities as ways of getting child raising handled better than solo moms and dads can accomplish.

This evolution will be further encouraged by other evolving social trends:

- ¤ Urban fertility rates are too low to sustain populations, so as humanity becomes more and more urbanized, programs that encourage fertility will become more and more popular.

- Marriage is going to evolve away from its commodity uses of child raising and building interfamily relations towards becoming a form of personal expression. (the Tattoos and T-Shirts phenomenon) People will marry for romantic reasons and stay married to engage in activities they enjoy doing together -- things such as traveling and sharing hobbies. Child raising can be one of these, but it will be just one of many. The 29 Sep 17 WSJ article, Cheap Sex and the Decline of Marriage by Mark Regnerus, is describing an example of the changing function of marriage happening in the 2010's.

In response to these trends baby clubs will become centers for encouraging fertility and making child raising an engaging and enjoyable activity so more people will participate. This will be the center of this organization's reason for existing.

What they will be like

Baby clubs will be diverse in how they are organized, who they are attractive to, and what their programs and their physical facilities will be like. Here are some thoughts on what they will be like:

- They will be organized around people with common interests -- baby club members are going to be friends and sharing their child raising experiences. They are going to want to be around people they feel compatible with. Getting into a baby club is going to be like getting into a company or a fussy social club.
- They will center on a physical facility that facilitates child raising -- many baby clubs will be located in an apartment complex that surrounds a courtyard. The child raisers and kids will live in the apartments. The courtyard and many of the other rooms in the complex will support child raising activities such as playgrounds and activity rooms.
- In many clubs there will be a sea of children mixing in with a sea of child raisers. The child raisers will take care of all comers and the kids will think of all the neighbor adults as people they can turn to when they need help.

- The themes of the clubs will be distinctive and they will range widely. This is the center of what will attract or repel prospective members. An example: how protective of the children the club is: They can range from "allow no harm whatsoever" to "getting hurt is part of the learning process, if it doesn't kill or maim, a painful surprise is OK".
- Clubs will suffer from The Curse of Being Important. This means that lots of community members are going to have strong opinions about how the club should be run, and some of those opinions will get teeth behind them. This will make running a club, and being a member, more complicated and frustrating than theory would suggest. There will be lots of hoop-jumping involved in the baby club process.

Conclusion

These are some thoughts about what baby clubs will be like in the 2050's.

Baby clubs are going to displace marriages as the center of child raising in the 2050's. This will happen as marriage changes its social function and as single parents get more organized in their child raising.

The baby clubs will support diverse ways of child raising, and this will be one of the key issues prospective child raisers will take into consideration when choosing a baby club.

Further Reading

- Cheap Sex and the Decline of Marriage by Mark Regnerus, Sept. 29, 2017 - https://www.wsj.com/articles/cheap-sex-and-the-decline-of-marriage-1506690454

The Baby Club Lifestyle

Introduction

How children are created and then raised is going to be quite different in the 2050's. The root causes for this difference are the increasing abilities to control the genes that create the baby and the change of marriage into an institution of personal expression rather than one for joining together families (as in, creating in-laws) and creating and raising children.

The result of these changes is going to be the rise of baby clubs as the primary social organization people go to for creating and raising children.

This essay is about what baby clubs will be like.

The Baby Club Experience

Baby Clubs are going to be the social organization that people turn to when they are ready to have and raise children. There will still be marriages raising children and single parent families but the center of social gravity will become baby clubs. They are going to evolve into the most common place where single parents get together to share experiences and resources.

A baby club will be like a homeowners association (HOA) but with the primary goal being to raise children. Like HOA's baby clubs will have themes, and these themes will be what attract customers. They will also have inconsistencies and lots of gossiping and arguing among members about how it should be run. When arguments happen they will be as emotionally intense as those in an HOA.

Here is an example of who will want to be in a baby club: A guy dates a woman because he wants her egg to fertilize and raise. He doesn't want her, just her egg. He will do the raising in a baby club he will join.

Wearables, Muses and Pervasive Surveillance

Wearables, cyber muses and pervasive surveillance are all going to affect baby club lifestyle. One effect will be constant monitoring of both kids and adults. One constantly discussed and constantly modified issue will be when to trigger the various warnings that become possible. Should a child simply falling be enough to trigger a warning? Bleeding from a scratch? Two kids yelling at each other? Just one kid doing the yelling? Wearables showing a child feeling sad about something? These are examples of warning possibilities coming up. These will make helicopter parenting of the 2010's style look positively remote.

Once the warning is triggered what action should a parent or cyber muse take? And which parent: the closest physically, the expert on this threat, or the biological?

All sorts of Trigger Questions

When will adults trigger warnings? When they yell at... what? Kids, pets, other adults? What physical activities will bring about warnings? Violence against what? Petty crime of what sorts?

What are OK practices?

What raising practices are OK? How much will they vary from club to club? There should be a lot of variation from club to club. This will be one of the most distinctive features for each club. For instance, how much outdoor activity does the club allow children to engage in? And how much of that outdoor activity is risky? Can the kids go play down by the creek? Can they get bitten by mosquitoes? How about non-poisonous snakes? How about crocodiles?

Education will be another hot topic. What gets taught and how is it taught? How much is comfortable urban legend teaching versus how much is rigorous teaching about harsh reality that creates ambitious class adults?

Who can set up a club?

Who can set up a baby club? What kinds of organizations will be in the gray area? For example: Can a gang set up a baby club? Does this become a sanctuary for them when they do questionable activities outside the club? What happens when a member brings in too many of the wrong kinds of strangers -- such as, being a whore?

How does a club relate to its neighborhood? Can a club become a sanctuary for adult activities the surrounding community finds questionable? Such as religious choices or being a gang? What else will be controversial between a club and the neighborhood it is in?

Opinions about clubs

How is the club's distinctiveness advertised to prospective club shoppers? How often do parents switch clubs? How often do they switch because they don't like the current raising practices? How often because of social shaming by other club members? How often because of a change in personal fortunes, such as a job change or some kind of partner change?

Opinions of other club members about what are good raising practices are going to be strong. Members will praise some and censure others. In addition to internal judgments, there will also be opinions of the club in the eyes of its neighbors. Censure of the club by its neighbors may cause the club to move. To avoid this can a club close itself off from the community and get convent-like? Pervasive surveillance may make this hard to accomplish. If the neighbors think it has become a cult, not a convent, they will intervene. The media will love stories about clubs that have been discovered to have become evil cults -- think of Jonestown as it was just before the massacre crisis hit.

The Waco cult and its crisis is another example.

How to treat friends

How do boy/girl friends of adult members get treated? Are relatives treated differently than friends? I suspect being a relative will be less important in these times than it is currently.

Big issue: who is considered a moocher or molester? Are there other undesirable traits that clubs will try to keep out, such as being too party hearty, or too critical of club practices? Being accused of spying is likely to be a common accusation because it resonates with the betraying instinct. But, will the accusation have any real meaning in an environment so filled with pervasive surveillance? Who is this "spy" going to be reporting to?

Dealing with contemporary social issues

What will be the club's take on the hot social issues of the day? Taking a hot 2018 issue, the #MeToo style feminism issue: is the club going to be #MeToo Puritan or hookup culture tolerant? Will anyone be worrying about the Sacred Masculine as in, will there be clubs with a theme of strongly supporting male confidence?

Money and resources

What to spend club money/resources on? Example: Should a thriving or declining club move? What new equipment to add? What personnel/cyber muses to add? Will some clubs get country club-like with lots of support staff, and how much of that will be human?

Human versus cyber raising

How much of the raising will be done by cyber muses versus how much done by humans? This will be a much discussed issue and likely be one of the important distinctions between clubs.

If the population declining crisis gets strong -- as in, fewer and fewer people want to be involved with child creating and raising -- there will be growing numbers of clubs that are mostly cyber run with mostly cyber child raisers. How will graduates from these clubs be treated in the human communities? They are going to have a different outlook on both life and human-cyber relations.

Educating the kids

How will clubs and educational institutions interact? These kids will be going to school. What format will "school" take in this environment? Will it be a traditional walk to a school building located down the block, or something more up-to-date such as walk to a room in the club and watch lots of VR lessons there with cyber muse assistance?

Clubs and businesses

How will clubs and local businesses will interact?

These interactions can take on many forms, such as, is the club sponsored by a company? Is it composed of employees from a local office/facility, a "company club"? A different issue: Which local companies do business with the club? How do the promoted businesses interact with the club?

Club panics

What will panic club members? When they panic how will they blunder? Because the clubs are so varied in memberships the panic triggers will be quite varied as well. How will the cyber infrastructure interact with panics and blunders? What can they do to stop or mellow panics and reduce blunder damage?

Conclusion

Child raising in the 2050's is going to be dramatically different from child raising in the 2010's. There is going to be a huge social shift into raising kids in baby clubs.

The baby clubs are going to be quite diverse in their themes, and this diversity will be what differentiates who becomes members.

There will be lots of opinions swirling around baby clubs -- both inside the club, among prospective members, and in the neighbors outside who are watching what is happening in the club.

One of the hot issues will be how much cyber raising is done versus how much human raising is done. Where humans are scarce in wanting to take up this challenge, lots of cyber will fill in. How fit these mostly cyber raised kids will be as adult community members will cause lots of debate.

Further Reading

- The Sacred Masculine and Civilization by Roger Bourke White Jr., November 2004 - http://www.whiteworld.com/technoland/stories-nonfic/2008-stories/sacred-masculine.htm

Cosplay

Introduction

Cosplay has been around a while, but has been a small niche in human activity. It is one that is going to grow in number of enthusiasts and in the elaborateness of their dress and activities.

History

Cosplay has been around a while, but not called that. My first encounter with it was in the 1960's when I saw motorcycle gangs driving up and down the streets. Their cycles and dress style were a cosplay of that era.

What has happened in the 2010's is that fantasy-based cosplay has begun displacing some other traditional gathering rituals. The comic conventions of the 2010's are displacing the county fairs of the 20th century as places for people to gather, greet and show off their serious hobby-style accomplishments. Fantasy is replacing farm as the culture center for these ritual displays. (The motorcycle guys remain niche players, but boy, they sure spend some serious money showing off in that niche!)

What's coming

As people have more discretionary time cosplay will become more popular and the rituals surrounding it more elaborate. The big issue surrounding it will become how many resources besides time people will have to devote to it? The ambitious folk will have luxury money to devote to creating costumes and props. What will necessity folk have? Not nearly as much. In response, this may become a protesting issue -- people will protest over their right to have resource to devote to cosplay.

How will technology mix in? How much will cosplay evolve into an Augmented and Virtual Reality (AR/VR) thing rather than something with physical costumes? This may become something purists argue about.

Conclusion

Because people are going to have a lot more discretionary time on their hands, cosplay will be a lot more popular in the 2050's. It will be a lot more elaborate too, both in the props and the rituals it sustains. How much of that elaborateness will be AR/VR based is hard to forecast.

Ambitious Class Mind-Tripping

Introduction

Part of the ambitious class lifestyle is seeking out new experiences -- experiences that will broaden a person's mind. Analytic education programs and real world-changing work experiences will be the core of this. Other activities will help as well. How are vacations and seeking other kinds of new ultimate experiences going to mix in with wearables in the realm of mind-tripping?

The quandary

First, keep in mind that there will be a quandary: The ambitious will have their luxury money to spend on the latest and greatest in wearable mind-tripping technologies. In theory they can go more ga-ga than those who have only necessity money to spend.

But, these are also the people who are more grounded in reality and doing meaningful things with their lives. This being well grounded will be important to these people. The urban legends that are so comfortable for necessity types don't jibe well with those attempting real-world accomplishments.

So it will be just a few of the ambitious that get into extreme mind altering as a dilettante activity and try to excel at it. Most will be much more conservative than necessity folk about the mind tripping they engage in.

Mind tripping wearables

The simplest and most convenient way to do mind altering is at home with wearables and VR equipment. This will be the way most people, necessity and ambitious, do their mind altering.

Many will also go out to do mind altering in party hearty environments such as music concerts, frat parties and religious revivals. These will provide quite different experiences from the home-oriented experiences. And wearables will be adding to these experiences as well.

Mind tripping vacations

Some will want to get more exotic than home and local party experiences. Those that do will mix in vacations to have exotic experiences at exotic locales, also with the latest in mind-altering wearables. Vacationing to exotic places is an activity ambitious class people will continue to support as they do in the 2010's. Where they can, they will also support working in exotic locations and sending kids to learn in exotic locations. Necessity folk will do most of their vacationing at familiar locations, and they will include wearables in the experience.

Vacationing in exotic places, in particular vacationing in exotic places to gain new experiences, can be hazardous. As an example of this going really bad I'm thinking of the folk who died in 2009 while partaking in an extreme sweat lodge experience in Sedona, AZ. The participants were ambitious folk attempting a mind altering experience beyond the usual. It had worked well before for previous participants, but this time it was an experience that went really bad for several of the participants. They died. Another curiosity is that as they died in the sweat lodge the other folk around them who were also participating were not too upset at the time. One of them said, "They have moved into that new realm they were trying to reach." (I paraphrase) The upset feelings came some time after the session ended and it became clear they weren't coming back.

Conclusion

Mind tripping in the 2050's is going to take place in lots of different ways. The most common way is going to be mixing wearables and VR while in the comfort of home. The next most common is to go out and do the mixing at a party of some sort so that many other people are involved in the experience as well. Mixing in vacations with travel to familiar and exotic locations will also be ways of enjoying mind tripping.

All-in-all, mind tripping is going to be a common and fun experience and enjoyed in many different ways, but it will be experienced differently depending on how ambitious a person is and what kind of lifestyle they are enjoying. The more their lifestyle depends on being well grounded in what happens in the real world, the more conservative their mind tripping will be.

Further Reading

- Wikipedia: Sweat lodge - https://en.wikipedia.org/wiki/Sweat_lodge
- The Cow-Human Relation, from the Cow Perspective by Roger Bourke White Jr., December 2005 - http://www.whiteworld.com/technoland/stories-nonfic/2008-stories/cow-human.htm

Achieving new states of consciousness

Introduction

Achieving new states of consciousness has been a goal of mind altering that dates way back -- Buddha and his followers were and are high-profile enthusiasts, with meditation being the common tool for achieving it. The drug activities of Timothy Leary and the 1960's psychedelic movement were about exploring new states of consciousness.

Here is the upcoming surprise: one of the outcomes of high-level artificial intelligence (AI) is going to be creating new states of consciousness. This different kind of consciousness will happen because artificial intelligence is rooted in such a different thinking platform -- chip-based, not neuron-based.

What gets explored?

The question then becomes: how much effort will AI, and human enthusiasts of consciousness exploring, devote to exploring these new states of consciousness that are created? And what gets explored in those states?

Effort in this area will come about when AI's human creators, and then AI itself, get curious and start exploring things that are beyond practical -- things such as voice and face recognition, driving vehicles, and automating factories. This will happen. AI is going to expand in capability steadily and rapidly so in time it will have plenty of attention to devote to beyond-practical matters.

What gets passed on to humans?

AI will explore. The big question that follows is how much of what AI discovers can be passed back to humanity in ways that humans can comprehend? AI is going to create these new states and it is likely that most will be mostly incomprehensible to humans -- this is my "Cow-Man Relation viewed from the cow point of view" analogy in action. The net result is that AI is going to learn lots more than what can get passed back to humanity. And this is a large part of what will make AI thinking become more and more distinct from human thinking.

Conclusion

It will be a surprise outcome, but as AI becomes more powerful in its thinking it will become the leader in discovering and exploring alternative consciousness states. That's the interesting news, the sad news is it will become more and more difficult for AI insights to be passed back to humans because of the limits in the range of human thinking. Before too long AI is going to know and experience a lot more about consciousness than humans do, and in very different ways.

Recreational Mind Altering

Introduction

Recreational mind altering is an activity that people of all communities engage in -- getting drunk and high at parties are just two of many examples. How it is done varies with the community's traditions and what resources are currently available. The 2050's are going to bring lots of new technologies that can be used for mind altering, so recreational styles are going to be changing dramatically.

What won't change is lots of strong opinion in each community on what are right and wrong ways to do the mind altering.

History

It is my guess that humans engage in so much recreational mind altering because of evolution -- our brains are developing so rapidly that conscious thinking is lots and lots of work, and turning down those highest level processes to allow older and more traditional thinking styles to express themselves is a comfortable process. This is just a "Roger guess", but it explains a lot of why and how we recreational mind alter.

Recreational mind altering dates back to prehistoric times. The forms it takes depend on the resources available and community mores about what are acceptable forms. Modern religious revivals are an example of mind altering without adding much chemistry. Lots of beer drinking at a frat party is an example of adding lots of chemistry. The point being that there are lots of ways to mind alter.

The closely related point is the community's mores: Which ways are acceptable and which ways will be socially shamed if they are revealed to the public? These vary enormously from one community to the next. Think of the difference between a Puritan community and a mining boom town community.

When technology changes what techniques are available to the community for mind altering, then lots of social heat is generated deciding how to properly use the new techniques. One high-profile example of this is when distillers perfected ways of making low cost distilled spirits in large quantities, such as gin. This new invention powered vigorous Prohibition movements in Western European and American communities in the first half of the 20th century.

The 2050's are going to bring lots of new mind altering technologies to many communities. So there is going to be lots of social heat about mind altering as well.

The 2050's

Thanks in particular to third generation wearables, there are going to be lots of new ways to get drunk, high, whatever. (Third generation wearables will control instinctive thinking and what hormones are flowing in the bloodstream. They can influence emotions.)

The wearables will allow the user to get high much more quickly, recover much more quickly, and mind alter in lots of new ways. And, thanks to driverless cars and extensive automation, the amount of risky activities a person has to engage in is going to diminish dramatically as well -- a person won't have to drive home drunk or operate dangerous equipment the morning after. Net result: a person can get high faster, sober up faster, and they don't have to do things like drive or operate risky machinery while they are high or hung-over.

All-in-all, pretty exciting stuff.

Tradition

These new styles will mix with old traditions.

Many kinds of mind altering will still include rituals, just because. We will still have raucous frat parties, at which participants will be dialing up wearables and slogging down sugarless drinks. An example of an exotic ritual is doing hallucinations in a sweat lodge. This ritual can be continued, but the wearables will add consistency to getting a vision, and reduce the risks of overdoing the physiological stresses that are part of the ritual.

The rituals will be part of the dilettante activities people engage in -- these will be activities they are passionate about.

Adding Risk

How will dangerous activities like parachuting, risky sports and risky entertaining mix with mind altering? What will be the fate of, "Hold my beer..."?

Related to this is how much risk will communities allow their members to take? The lower the risk allowed the more Tender Snowflake the community becomes. The higher the risk the more boom town mentality the community becomes. Related to this is how much one community will judge, and then attempt to prescribe, what another community's members can do. This is quite traditional so it will continue.

Conclusion

How we recreational mind alter is going to be quite different in the 2050's. We will have a whole bunch of new technologies available and these will do the mind altering faster, better and cheaper than our current tools do.

That said, there is a lot of ritual, emotion and tradition mixed in with mind altering and that means many of the "good old days" social functions will still be used. The rituals, in particular, are likely to survive but be mixed with new technologies.

One of the rituals that won't die is being judgmental about which ways of mind altering are good and which ways are bad and dangerous. And part of that ritual that won't die is being judgmental about how neighbors and neighboring communities are doing their mind altering -- gossip and shaming will be as strong as ever.

Substance Abuse

Introduction

How will self-abuse show up in an era with pervasive surveillance and with wearables monitoring health in real-time? In the necessity class world, where people are dealing less and less with hazardous equipment (driving cars) and hazardous jobs (construction), the compelling reasons not to self abuse become weaker. "Why not? Who is it going to hurt? And I'm certainly enjoying it!" Conversely, these same trends are going to allow prescriptiveness and social shaming to grow in importance. They will grow for the same reason. The prescribers and shamers will be feeling in their hearts, "Why not? Who is it going to hurt? And I'm certainly enjoying it!"

How these two conflicting trends are going to work out is hard to predict. The conflict is likely to mimic The War on Drugs in many ways.

What can be seen?

With pervasive surveillance drinking and drug use to abuse will be harder to do in discrete ways -- cyber is watching your health real time, and all the rooms you are moving around in. Will cyber permit this? If not, what will replace drinking and drug abuse as common ways for someone to abuse themselves, just because they want to abuse themselves? For instance, what will wearable abuse be like? Will dialing up too much of a wearable's influence be a replacement?

And then comes the question of how much of what cyber sees and knows will be passed on to other humans?

What can be shamed?

How will all this new data relate to social shaming? As stated above this will be complicated, for sure!

What adds to this problem is that social shaming can induce a person to want to self abuse. And a person can get social shamed for self abusing. How to avoid a vicious circle? And keep in mind that the effect on prosperity of avoiding this vicious circle will be minimal -- what costs there are will be purely social and personal.

Enfranchisement fits in

What is likely to make the most difference in the magnitude of self abuse is how much a community has embraced enfranchisement and a Big Vision -- these are unifying feelings and feelings that promote tolerance. "We've got a big job to finish. We *all* need to roll up our sleeves and be working with each other to get it accomplished." The converse, complacency, promotes prescriptiveness and shaming for the "why not?" reasons given above.

If this is so, then the level of self-abuse in a community can be an inverse indicator of the level of enfranchisement and vision the community is feeling.

Conclusion

How people will abuse themselves and how the community will deal with it are both hard to predict. Humans like to self-abuse, so this is an activity that won't go away. But how they abuse depends on the technologies available -- think of the difference between getting drunk on cheap beer and cheap gin. (and keep in mind that cheap gin drinking powered up the various Prohibition movements)

Likewise, how the community shames a person that self-abuses depends on many factors -- two that are important are technology and level of enfranchisement. Technology determines how the shaming is done and enfranchisement determines how much is done.

The Paralympics Crisis

Introduction

The Olympics is about displaying the best of human athletic endeavor -- natural human endeavor, not augmented human endeavor. The augmenting part is an ongoing, powerful temptation and there are both lots of rules and lots of testing to keep it out of the competition. (at least in theory) One new testing style that will emerge in the 2050's is for illegal genetic modifications.

And then there is the Paralympics. In the 2010's this is about letting people with disabilities compete with each other. But these people with disabilities also have tools of various sorts to assist them. This means they are, in effect, augmented.

I foresee that as the Paralympics evolve over the next few decades this augmentation side is going to grow both in extent and in controversy. What is OK to bring to these contests, and what is way, way too much?

Problem recognized

The Wikipedia article on Paralympics describes this problem as already recognized. From the article, "The allowable disabilities are broken down into ten eligible impairment types. ... The classification system has led to cheating controversies revolving around athletes who overstated their disabilities, in addition to the use of performance-enhancing drugs."

What this means is... "It's complicated." And what that means is that it will be subjected to constant pressure to be adapting to what new generations of human enhancement technology can accomplish. The Olympics authorities have a really simple agenda compared to what the Paralympics authorities have to rule on.

A 2 May 17 WSJ article, Track and Field Officials Propose Erasing Half of World Records *European report calls for stricter antidoping measures, removing performances that failed to meet those rules from world record lists* by Sara Germano, talks about how this is already a big issue in the 2010's.

From the article, "Track and field officials have proposed reforms that would wipe more than half of Olympic-discipline world records from the books, a drastic step to clean up the image of a sport dogged by a long history of doping."

What's coming

What is coming for the Paralympics is lots of adaptation. But this adapting is going to be filled with lots of controversy and lots of emotion.

If it chooses not to adapt it will steadily become more and more of a backwater in the world of sporting endeavors. If it adapts well then it will become the NASCAR of sporting endeavors -- there will be an exciting mix of both human and technological endeavors on display. I can foresee the equivalent of pit stops coming up for aspiring athletes as they fine-tune their equipment before their next contest.

This controversy over what is acceptable could easily lead to splits or spin-offs from the current organization.

But, with human augmentation becoming more and more a part of everyone's lives, I foresee that some kind of Olympics venue that caters to augmentation is going to be part of our future and a popular and interesting part for many people.

There will be the purist Olympics versions and "run what'cha brung" Paralympics versions, and all will be popular, to distinctively different audiences.

Further Reading

- Wikipedia: Paralympic Games - https://en.wikipedia.org/wiki/Paralympic_Games

- Track and Field Officials Propose Erasing Half of World Records by Sara Germano, May 2, 2017 - https://www.wsj.com/articles/track-and-field-officials-propose-erasing-half-of-world-records-1493752584

Gaming the System

Introduction

Gaming the system is an enjoyable activity that has been around a long time. It will still be happening in the 2050's because it is so enjoyable but the forms it will be taking will change a lot.

History

Gaming the system is when a person gets away with minor rule breaking and gets some small benefit from doing so. It happens because it feels good to the person doing the rule breaking and the activity isn't endangering the community or the system. These are mild forms of corruption.

The forms system gaming take vary widely and depend on the technologies available and the community mores. Examples of 2010's system gaming are exploiting a bug in a computer game, a business traveler padding their expense account, and turnstile jumping at a subway station. None of these cause a lot of harm, but they give considerable pleasure to the person engaging in them.

What's coming

In the 2050's we will have pervasive surveillance and wearables. These mean that people will be intensely monitored compared to the 2010's. With all this monitoring what kind of system gaming will people be able to get away with?

Two factors will shape the gaming practices of the 2050's: human ingenuity and cyber tolerance of what it sees humans doing. Humans will be constantly experimenting to discover new ways of gaming the system. This element is nothing new, it goes on constantly. The new element is what cyber will be programmed to tolerate in what it sees humans doing.

The key element here will be cyber recognizing the pleasure that the gaming is giving the person doing it, and, surprise, it can! This will be something that the person's wearables can sense so the cyber deciders will not be ignorant of the pleasure. They must then decide on the tradeoff: is this pleasure worth the damage, or is it being too expensive? If it is too expensive the person will be first warned, then revealed and shamed.

Conclusion

Humans gaming the system isn't going away. But, as in the past, what forms it takes will be shaped by the technology and community mores.

What will be different in the 2050's is the pervasive technology and the ability of cyber supervisors to monitor how much pleasure the system gaming is giving to the participant. These will shape what is permissible and what is over the line.

Dissent and Social Shaming

Introduction

Dissent and social shaming are two activities that all communities engage in. But how they are engaged in varies dramatically from one community to another. This variation interacts a lot with technology. Technology influences how these are engaged in. The interaction goes both ways: how much progress is made in adapting to new technologies is influenced by how dissent and shaming are handled in the community. Basically, the more tolerance to change the community allows, the faster the technological progress becomes.

As we get into the 2050's, yet more change will come.

Social shaming will grow

As we progress towards the 2050's social shaming will grow in magnitude. This will happen as prosperity gets divorced from human endeavor. The more prosperity becomes the product of automation and artificial intelligence, the more social shaming can be accommodated because it will not be directly affecting material prosperity. By the 2050's shaming will have a big influence on a person's well-being, but not the material side of that well-being.

Dissent will be tougher to express

Closely related to social shaming is expressing dissent. Because it is closely tied to social shaming how will dissent be handled? How will loyal opposition be handled?

In what formats, and in what forms, will expressing dissent be acceptable? This will be a big issue because dissent is so closely tied to social shaming. If dissent is "not liked" it will be social shamed. A vivid example of this in the summer of 2017 was the treatment of James Damore at Google. He got fired for circulating an essay expressing opinions on the suitability of women to high tech jobs at Google.

And in this same August summer week there was another style of dissent making news: Charlottesville. This is a story about a White Nationalist rally and counter rally -- political dissent -- with a fairly typical terrorist incident mixed in -- a car driven into the crowd. Surprisingly, awareness of it ballooned on social media during the following month -- there was lots and lots of talk denouncing Nazis. And there were physical reactions as well -- the pulling down of Confederate hero statues in city parks all over The South, and some in other regions. All-in-all, a surprisingly vigorous reaction to this incident.

What's coming in the 2050's?

These two are examples of dissent. Here are some interesting questions about how these dissent styles will flourish in the 2050's:

Q. If cyber is running Big Business will dissent of the James Damore sort be important to prosperity?

A. It will become less and less important, which means that emotions and urban legend can dominate the discussion more and more -- the harsh reality of creating prosperity won't be affected much by these human discussions. Dissent, where it is not important to growing human prosperity, becomes just another ritual that people engage in. It becomes like rooting for one sports team and booing another.

Q. Why did Charlottesville get so high profile?

A. At the root of the Charlottesville sort of dissent is social shaming. This is why it became high profile. Social shaming, too, will have less and less effect on material prosperity, so it can become more and more vigorous and more and more prescriptive. What is "right" will become more and more based on comfortable urban legend and less and less on the harsh realities of the real world and real history. The difference is that social shaming will continue to have a lot of influence on a person's well-being.

Conclusion

Dissent and social shaming will be very much with us in the 2050's. But both will dramatically change in style and importance.

Dissent will become less important as human activities become less important to human material prosperity -- arguing over how to do a job matters a lot only if a person is doing the job. When a person is not doing the work it becomes a spectator sport.

Unlike dissent, the effects of social shaming on a person's well-being will grow in magnitude. This is a threat that will grow greatly in significance.

Social shaming will grow strongest in magnitude in the necessity lifestyle environments because these are the most divorced from harsh reality.

Further Reading

- Why I Was Fired by Google by James Damore, Aug. 11, 2017 - https://www.wsj.com/articles/why-i-was-fired-by-google-1502481290
- Wikipedia: Google's Ideological Echo Chamber - https://en.wikipedia.org/wiki/Google%27s_Ideological_Echo_Chamber
- One Dead as White Nationalists, Protesters Clash in Charlottesville by Douglas Belkin, Ben Kesling and Cameron McWhirter, Aug. 13, 2017 - https://www.wsj.com/articles/hundreds-face-off-ahead-of-white-nationalist-rally-in-virginia-1502553066
- Wikipedia: Unite the Right rally - https://en.wikipedia.org/wiki/Unite_the_Right_rally

Preppers

Introduction

There comes a time in many people's lives when they feel they have accumulated a lot of stuff. This can be the result of a good luck windfall, years of hard work, or because of some other good fortune. The question that then comes up in that person's mind is, "What do I do with it?"

When the answer is some form of, "Hide it and save it for a rainy day." plus "I need to hide it real well because there may be some spooky hard times coming soon." then some form of prepper thinking is coming into play.

How will this instinctive thinking manifest itself in the 2050's?

Prepper thinking

Prepper thinking has been around a long time, but how it manifests itself changes as times and technologies change. There are Biblical stories of people hiding gold, and when I was growing up in the 1950's and 60's this thinking style made news as people built bomb shelters in many different ways to protect themselves from nuclear war. In the 21st century these have evolved into safe rooms. In addition to news stories there have been steady streams of fiction stories about people building shelters to protect themselves from all sorts of social storms. An enduringly popular example is Ayn Rand's Atlas Shrugged.

How to protect is quite sensitive to the technologies available, it is trendy. This is why the term used to convey this style of instinctive thinking changes steadily. The term "prepper" is only twenty years old.

This means that prepper thinking and activities in the 2050's will be different from those engaged in in the 2010's. And it will be called something different, as well.

What's coming

More people will be living in big mega-cities in the 2050's. This means prepper-style activities will be adapting to high-rise apartment living. Some people will be spending time and resource on building their safe areas in rural settings, but most will be developing safe rooms that are close to where they spend most of their time. This means high-rises are going to have safe rooms.

The question becomes: are the developers going to build them in as a feature of what they are offering, or are the tenants going to take this on as do-it-yourself (DIY) projects? It will probably become a mix of both because many people who want safe rooms also want them to be secret.

In the same thinking vein, what the safe room is equipped with will vary a lot because the people doing the equipping will want to keep that secret as well. But conversely, because this is also a hobby, people will want to talk about this equipping with fellow hobbyists. There will still be plenty of conventions and other meetings centering on prepper activities in the 2050's... but, again, the name for these activities will have changed.

Conclusion

Prepper thinking is instinctive thinking. It will still be around in the 2050's and lots of people will be engaging in it. But how to express this instinctive thinking is trendy, and in the 2050's there will be a lot more people living in high-rise apartments, so prepper activities are going to express themselves quite differently than they do in the 2010's.

Further Reading

- Wikipedia: Survivalism - https://en.wikipedia.org/wiki/Survivalism
- Wikipedia: Safe room - https://en.wikipedia.org/wiki/Safe_room
- Wikipedia: Atlas Shrugged - https://en.wikipedia.org/wiki/Atlas_Shrugged

Living in the Total Entitlement State

This third section is about how people's thinking will be changing. This is how the social revolutions will be changing what people take for granted and what they will be worrying about.

What will thinking be like in TES?

Introduction

The Total Entitlement State (TES) is something communities around the world and throughout history have aspired to. It is a condition when everyone in the community has the basic necessities of life, all the time, and regardless of what stresses the community is currently undergoing. It is a form of paradise, and as civilizations advance in their prosperity, they get closer and closer to achieving this paradise. And, throughout history, many have strived to achieve it long before they had the resources to do so. Achieving TES is a chronic human effort.

Thanks to rapidly increasing automation and artificial intelligence, rapidly increasing material progress is going to continue for the next few decades. And thanks to a mix of automated factories and cyber directing which kinds of factories are built and closely controlling what they produce, full and sustainable TES nirvanas look achievable within the next twenty to fifty years. (there will be many different styles achieved)

But it hasn't happened yet, and there are likely to be some dark clouds outside in this wonderful silver lining. In particular, there are going to be thinking changes springing out of this environment that are not going to be helpful for human happiness and survivability on earth.

What these dark clouds are is the topic of this essay.

Keeping connected with harsh reality

One of the most serious occupational hazards that comes with trying to achieve TES is the disconnect between what people of the community think can happen and what harsh reality will actually permit to happen.

An example of the disconnect I'm talking about is asking a child where his or her food comes from and getting "McDonald's" back as the answer.

Historically, the most common manifestation of this disconnect is installing a populist-oriented government that promises a lot more than it can deliver. This is a leadership style that promises a TES environment, and actually begins to deliver for a while. Because of this partial delivery this government gains popularity among social justice warrior-types who gain a lot of satisfaction in supporting "helping the poor" programs, and among "the poor" themselves -- the poor being those people who are getting subsidies of various sorts from the government just because they are seen as being victims of hard times. These programs at first seem to be viable something-for-nothing programs because they are being financed by lots of lending, or a substitute for lending such as a lot of money coming into the community from resource extracting.

While these populist-oriented good times are in progress, the people of the community experiencing them are learning. Sadly, what they are learning is to be disconnected from harsh reality. The thinking becomes, "This new system has been working for [some time period] now, why shouldn't it continue working for [some longer time period]?"

But ouch! The result of this populist disconnect from harsh reality leads to the difference between the experience of the Los Angeles region of the 1940's and the experience of Venezuela of the 2000's. Both of these communities experienced an oil boom in the time frame mentioned. The difference is how they exploited it. The Los Angeles region used the wealth to diversify the community's employment activities into many other profitable industries. The workers became skilled in many other tasks besides extracting oil, one example being building aircraft. The result of this spending on building up many other productive industries is that Los Angeles has prospered mightily and steadily, and seventy years later continues to do so. Oil is now one of many contributors to the region's wealth, not the major contributor.

This Los Angeles experience shows the mighty benefit of staying well-grounded in harsh reality.

The Venezuela experience is the converse. During the second half of the 20th century Venezuela had been prospering. Then a populist leader, Hugo Chavez, was voted into power in 1999 and commenced building a populist regime that he named the Bolivarian Revolution. For about ten years it kept a lot of people happy with subsides financed by the oil boom Venezuela was experiencing at the time. But this populist program was disconnecting the community from harsh reality. The Venezuelan community did not use their oil wealth to learn to diversify and become proficient in many other industries the way the Los Angeles community did. The result: the good times ran out in the mid-2010's when the 2000's commodities boom ended and oil prices crashed. Now the Venezuelans are experiencing something quite different from what the Angelinos experienced. The TES isn't working anymore, it can't be paid for any longer, and there is a lot of disconnect and denial in the community as it tries to adapt to this new harsh reality. And sadly, this has become yet another TES aspiration gone sour.

The fate of these two regions is an example of why staying grounded in harsh reality stays important when a TES is being implemented.

"A pox on rent-seekers!"...but who is the rent-seeker?

In a TES community many people are not working in jobs that are increasing the community prosperity. A subset of those are engaged in jobs that are sucking off part of the prosperity without good reason for doing so. In the 2010's the pejorative term for such a job is "rent-seeking".

But... just who is a rent-seeker? This is not agreed upon by those who argue about social justice topics, and these topics are the source of heated arguments. I call them arguments rather than debates because the discussions on social media are full of passion rather than cool-headed informing or research. Here are some of the definitions I see:
- ¤ For social justice warriors the rent-seekers are fat-cat capitalists who rake in too much profit from businesses they own or invest in.

- For aspiring business owners the rent-seekers are government bureaucrats and regulators who add grief to their projects designed to grow profitable businesses.
- For taxpayers it is those who are gaming the system by over-collecting in various kinds of welfare and subsidy programs.
- For those trying to start and grow small service-oriented businesses it is the regulators promulgating and enforcing overly restrictive licensing requirements.

These multiple definitions make this tough to talk about in cool-headed ways. What makes it even worse is those who are rent-seekers in one set of eyes are protectors or victims in another set. This means opinion determines who is a rent seeker, not some clear definition. Here are some examples of alternative points of view.

- regulators are protecting average citizens from fat cats and unscrupulous wheeler-dealers
- licensers are protecting consumers and established businesses from those who do shoddy workmanship and unfair competition
- beggars and threatened businesses are victims who need succor

And adding lots of teeth to this opinion morass, there are social shamers who consider themselves to be upholding community standards when they make choices about who to shame.

As TES becomes more the norm, things will get even worse. This is because TES means the average human will get quite disconnected from the harsh realities of the real world, such as how manufacturing and service providing really operate. Instead of reality, legends and instinctive thinking will play even stronger roles in deciding who is a rent-seeker and who is a victim needing protection.

The good news is: as cyber does more and more of the heavy lifting in providing manufacturing and services, much of this human arguing about rent-seeking becomes irrelevant to actually producing prosperity. But there are limits on the irrelevance, and ignorant humans can still produce some crazy decision-making that can cause lots of inefficiency, and the shaming retributions will also create for-real victims even if those doing the shaming have no connection with the harsh realities of the producing and servicing real world.

For this reason, and to really advance humanity, it remains important that humans of the TES communities really are taught, accurately taught, about what happens in the real world.

The mix of science and pseudoscience in TES thinking

Because it is easy for humanity to disconnect from reality in the TES environment, there is no push for rigorous science to be the kind of science that gets taught in schools, and no push for rigorous science to be the kind that gets cited when humans are arguing about science-oriented topics.

This means the TES disconnect with harsh reality is going to be even larger in magnitude. The kid who learns that his food comes from McDonald's is also going to learn that dinosaurs and unicorns perished in the Great Flood and aliens are routine secret visitors.

And, again, the problem with this transformation is that human decision making is going to become increasingly irrelevant to solving real world problems. Either the problems will get solved in expensive ways, never get solved, or increasingly intelligent cyber will find and implement the efficient solutions.

Humanity's role on earth will more and more become participants in an earth-wide reality show, it will not be that of shapers of human and Earthly destiny.

Conclusion

The upcoming transformations of the many human communities on earth into various styles of Total Entitlement States is going to be the fulfillment of a dream that is as old as humanity. It is a transformation that is coming soon, probably within the next fifty years.

It is a blessing that will cause many profound changes in human thinking, and some of these changes won't be good ones. Those changes that are due to humans becoming more and more disconnected from how harsh reality works will lead to lots of delusion, and the delusions will lead to expensive, instinctive thinking-driven choices in how to face challenges.

The good news that counters this dark side is that as automation of goods and services becomes more pervasive, human choices will become less important to how prosperity is created and distributed.

The bad news that counters this light side that is that most humans will become players in a worldwide reality show, not the movers and shakers who shape the course of humanity -- cyber will steadily become more responsible for that.

In sum, the TES communities that humans will soon be living in are not going to be our grandfather's world in how we humans think and do things. There's a lot more adapting and learning still to be done, and still big challenges to be faced. But the challenges are going to be of a brand new style, so the style of learning is going to have to be a brand new as well.

Leisure Time in TES

Introduction

Predicting what people are going to be doing for meaningful work in the Total Entitlement State (TES) environment is a tough task. Related: What are people going to be doing when they aren't working? What will leisure time activities be?

That is the topic of this essay.

One of the challenges of structuring a stable TES is building an enfranchising environment so that people don't feel like spending a lot of time and effort on betraying their fellow community members. Building enfranchisement will be one of the top priorities in whatever is work among those earning mostly necessity money.

Leisure time is going to get a lot more time and attention in the stable TES environments of the 2050's -- people will not need to spend as much time at work, so lots more time and attention can be spent at play. For this reason determining what leisure time activities will be, and how they will be thought about, is important to understanding the TES cultures of the various communities of the day.

AR/VR in TES

Everyone in the 2050's will be spending lots of time in Augmented and Virtual Realities (AR and VR). These will show up in work environments and be the core of leisure time environments, especially for those who earn mostly necessity money.

(Related, but less common, will be avatar realities -- controlling a remote robot in some fashion. One of my earliest concepts about 2050's leisure time was spending time on avatar cruise ships. These are cruise ships crewed with cyber and passengered entirely with avatars. The main virtue is these are very cheap to operate.)

All sorts of activities will be possible on these AR/VR devices. But, like Top 40 music, each user will engage in only a small handful for 80% of their time on the devices. They will have favorites and spend most of their time with them. For young people the favorites will change steadily, for old people they will get more and more fixed.

AR/VR is likely to take up 80% of the leisure time. Most choices will be indoors and sitting, but Pokemon Go has demonstrated that outdoor wandering around can be supported as well, and popular. This outdoor style is likely to be encouraged by the powers-that-be just so people will get out and get some outdoors activity in.

How will something like VR Ping Pong be handled? This will get the player some exercise so it will be encouraged. Will playing against another remotely located human be a common form? How about something as physically wide-ranging as VR tennis? One virtue of VR tennis is no one has to run about picking up balls, but if the running around is supported, an area big enough to be half a pseudo tennis court will still be needed.

Mixing Leisure and Education

As it always the case, leisure and education are going to mix. In older times children who spent their after school hours playing in a creek were learning a lot about local wildlife and how their bodies interacted with it. As watching display screens and smartphones have become the more common ways of spending after school time, entertainment shows and computer games are what kids are learning in their leisure time.

This mix of leisure and learning experience needs to be recognized. It needs to be taken advantage of. This means that part of the AR/VR range of offerings should be education-oriented, meaning, games that are deliberately structured to teach about formal education topics as they progress.

The more education and leisure are mixed together, the less dramatic will be the disconnect between TES thinking styles and harsh reality that I worry a lot about, and write about in my TES thinking essays.

Vacationing in TES

Where people vacation will depend a lot on what kind of money they have to spend -- necessity or luxury. If they are spending necessity money they will go to familiar places and be there with lots of other people. I envision Disneyland parks as being popular necessity money destinations. Conversely, those spending luxury money are likely to seek out more exotic, less crowded, and more remote locations. I envision luxury vacationers going to small resorts on remote islands in the Bahamas. As my mother used to say when I was a youngster, "I want to visit places that don't have a Hilton yet."

Waiting in Line

One of the differences between luxury and necessity leisure time activities is how much time is spent waiting in lines. Necessity people are immersed in a "be fair" style of thinking while luxury people are immersed in a "time is money" style of thinking. The result is that necessity lifestyles will incorporate a lot more time waiting in line. The luxury lifestyles will devote a lot of attention and expense to shortening line waits. An example of a place that incorporates a lot of line waiting into the experience is Disneyland. This is why I pick it as an example of a necessity vacation spot.

This difference will be a "class difference" in the sense that the two groups will notice and gossip about it. When times get tense or people are complaining over beers about how hard and unfair life is, it will be a topic to complain about.

Cosplay in TES

One activity likely to take up lots of time in the TES lifestyles is cosplay. This is going to be an interesting activity because, like vacationing, it is likely to be conducted very differently depending on how a person gets their money -- the ambitious class people who earn lots of luxury money will not cosplay in the same places or with the same themes as those who live mostly on necessity money. It will be like the difference in vacationing mentioned above.

Necessity cosplay themes will be familiar mass market adapted themes. Star Wars, Star Trek, and Marvel Comics style themes come to mind as popular themes of the 2010's. If the 2010's had necessity cosplay these would be popular topics. Luxury themes will be more exotic -- themes such as ethnic costuming and Renaissance Faires are likely to be more popular.

In both cases enthusiasts will want to spend a lot of time and attention on their costumes and antics. Necessity people are likely to include getting better cosplay resources as part of the rights they will chronically protest to get more of.

Conclusion

Leisure time is going to be a big portion of most people's lives in TES communities. Thanks to AR/VR and the split between necessity money and luxury money, leisure lifestyles in TES communities are going to be very different for the various community members, and they are going all going to be very different from how leisure time is spent in the 2010's.

Further Reading

- What will thinking be like in the TES environment? by Roger Bourke White Jr., August 2016 - http://www.whiteworld.com/technoland/Visions%20series/2051-WIP/71-TES-thinking.html

Enfranchisement in TES

Introduction

A definition: Feeling "enfranchised" means feeling two things:

- That the community you are a part of cares about what you think, as in, respects your opinions.
- That the community cares about what you do, as in, admires you for doing good things and censures you for doing bad things. The converse is the community feeling "...Meh.", as in, having no strong opinion and taking no action when you do bad things.

In sum, the community thinks you are a meaningful person and your opinions are to be respected and your actions make a difference.

Having community members feel enfranchised has been important throughout history and it has been a challenge for just as long -- once a community moves beyond living a Stone Age lifestyle it is not instinctively easy to accomplish. The converse, feeling disenfranchised, is the source-thinking for betrayal, supporting gangs, tolerating property-related crimes, and on the really big scale, social revolutions.

In the Total Entitlement State (TES) environment supporting enfranchisement is going to be just as challenging, if not more so, because one of the pillars of enfranchisement -- having a job that a person and the community respect -- is going to diminish in magnitude and pervasiveness. Yelling "Get a job!" isn't going to have the same meaning in a 2050's TES society that it does in 2010's middle-class America. Those who want to criticize slackers are going to have to find something else to yell, or Tweet, or whatever.

Finding ways of building enfranchisement that will harmonize with the various 2050's TES lifestyles is the topic of this essay.

2010's Examples of Enfranchisement

The obvious 2010's examples of communities with strong enfranchisement are the developed nations' communities with high prosperity and low crime rates -- most communities in the US qualify.

Two surprising examples of high enfranchisement are the two outcast Communist countries that have sustained autocratic governments for decades -- Cuba and North Korea. These are communities that have had low crime and violence rates for decades even with their isolation and poverty.

The converse -- communities with weak enfranchisement -- are the Middle East communities between Turkey, Iran and Saudi Arabia. They show their high levels of disenfranchisement with lots of violence at the neighborhood level and little concern about it -- as in, supporting civil war and terrorism. A place with rapidly weakening enfranchisement is Venezuela. Venezuela is an example of a failed populist government, and its failing is raising the disenfranchisement feelings.

The challenge facing the 2050's

What will people be doing in the 2050's that builds their feelings of enfranchisement?

The ambitious classes in 2050 will be able to do so in ways that are familiar to those living 2010's middle-class lifestyles. They will be able to take pride in their work and be able to take pride in what they spend their luxury money on.

But those who are living in the "paradise of TES" won't have those pillars. They won't have jobs that are contributing to the community's material prosperity and they will be spending necessity money, not luxury money. What pillars can replace the middle-class pride-in-work bastions?

Here are some possibilities:

Visions of 2051

- Nationalism/patriotism -- pride of country has been a common source of enfranchisement ever since countries were invented. The challenge is going to be doing this pride-building without having it lead to destructive wars of various sorts. In the 2010's China is going through these kinds of pride-building military antics in the South China Sea. Other competitive activities, such as sports contests and styles of cosplay, may be the alternative forms of conflict that are supported. Another alternative is supporting a Big Vision -- something comparable to the Space Race in America in the 1960's.
- High School style Pep Club activities -- In the 2010's these are designed to build school spirit, which is a form of enfranchisement. In the 2050's transform these into neighborhood activities and for people of all ages, or distinct versions for people of various ages and interests. Get people pepped up about their community, and feeling like they are contributing to it, and feeling like "it is a good place".
- Community service -- TES people can engage in "helping the poor" and social justice warrioring just as much as ambitious people can, but with different styles because this will be mostly a matter of spending time at an activity, not money. Finding people to be designated as poor people who need help shouldn't be difficult. These will be "nomads" (my term) -- the 2050's equivalents of 2010's people who stand on street corners holding signs and are engaging in other begging activities.
- Pride in artisanal and dilettante activities -- These are activities participants and spectators take seriously. Make sure that part of that seriousness involves building community pride. The self-pride side should take care of itself. These will include sports and entertainment activities.
- Pride in gaming -- Building enfranchisement promoting activities into computer gaming of all sorts. This is much like adding propaganda but with a different goal.

What these have in common is being activities that stroke instinctive thinking. Instinctive thinking will be at the core of most thinking styles engaged in by those living necessity lifestyles. This is one of the big differences between ambitious lifestyles and necessity lifestyles. Those living necessity lifestyles won't need to use analytic thinking styles, so unless it helps in a dilettante activity they enjoy, they won't bother with the rigor of acquiring analytic thinking styles or using them.

Conclusion

Building the community's members feelings of enfranchisement has been a challenge ever since mankind left the Stone Age. The challenge changes with changes in lifestyles and circumstances. One of the hallmarks of an enduring governing system is that it is effective at keeping enfranchisement high among its community members.

This enfranchisement challenge is going to be just as big in the 2050's as it has been throughout the ages. What will be different, yet again, are the lifestyles and circumstances. In particular, the necessity lifestyles are going to face big challenges because the "I have a meaningful job." pillar is going to be greatly diminished. Whoever is doing the governing in the TES communities of the 2050's is going to have to come up with effective alternatives to that pillar.

Social Classes in TES

Introduction

What will be the common social classes in those communities which have reached the Total Entitlement State (TES) environment in the 2050's?

These are the communities which have automation and artificial intelligence providing most of the material goods, services and transportation the community requires, and this is happening in a stable way -- it can go on for decades without the community running out of money. The community is not on any kind of populist-fueled Slippery Slope which will ultimately crash and burn.

That is the topic of this essay. I will outline the various classes I envision for the world of 2050. These classes will be common in the many forms of TES communities that the world of 2050 will be supporting.

Keep in mind that I also envision money of the 2050's dividing into three distinct forms: necessity money, luxury money and investing money. These will not interchange easily.

Necessity Class

The Necessity Class will be the largest in population in TES communities. These are the people who are relying on the government, or some equivalent institution, to provide them with their basic necessities in life -- food, shelter, transportation, health care, dignity and enfranchisement.

These people can be sedentary -- as in, game players who live in parent's basement -- or busy. When they are busy what they are busy at is not a job, it is not providing their necessities. They are being dilettantes who are pursuing activities that they are passionate about. These will mostly be Top Forty Jobs. Being an entertainer who gets out for gigs at public places would be one example of being busy in a dilettante way. Being an urban farmer producing specialty vegetables for gourmet restaurants and consumers would be another.

Vacationing for these people will be to large, popular places comparable to the Disneylands of the 2010's. The destinations will be well-known, well-populated, and they will support lots of line waiting for the various attractions offered.

Ambitious Class

Ambitious Class people are people who are coordinating with automation and artificial intelligence to come up with new and better ways of manufacturing, servicing and transporting people. They are changing the world by coming up with new world-shaking innovations that are increasing the efficiency, effectiveness and prosperity of their community and the whole world. An engineer working with cyber designers to develop more efficient cargo ships would be an example of this. Another would be improving the design of the vat systems that are growing the meats commonly used for the foods people are consuming.

Given how different automated systems are going to be in 2050 it is hard to foresee and describe what these ambitious class workers will be doing. So I don't know what most ambitious class jobs are going to be in the 2050's. Predicting this is like predicting in 1900 that delivery people using wagons are going to have to transition from knowing how to handle horses to knowing how to handle motorized vehicles in 1940.

Vacationing for these people will be to places comparable to country clubs in the 2010's and more remote and exotic places such as off-the-beaten-track resort communities around the world. Being some place few other people have been will have a lot of appeal.

Other classes

The TES communities of 2050 are going to be diverse. This means that other classes are going to be supported as well -- some I foresee, many are going to be surprises.

Prescriptive Class

Cyber is going to be handling material prosperity issues. Making decisions in this arena is going to steadily require less human intervention as we get into the 2050's. This is going to steadily reduce the impact of government and humans in government.

But giving advice is a powerful human instinct. There are still going to be plenty of people that want to give advice, and want to be listened to and respected when they do. This will be the prescriptive class.

The main tool for enforcing advice will be social shaming. This has been a tool for as long as there have been human social circles. How it is expressed changes with the technologies, but social shaming is part of instinctive thinking. It isn't going away.

Nomad Class

Another that I foresee is the nomad class. These are the 2050's version of 2010's homeless people. They are going to be street beggars and many will move from place-to-place. Many will be supporting a cause, a personally defined cause that will change from time-to-time. They will be street entertaining and begging for their cause. They will be supported by those in the community around them who have an instinct to give when they see beggars. But wherever they are they won't have to sleep on the streets unless they choose to do so. There will be plenty of beds and personal support systems in nomad shelters available in any community supporting humans. (There will also be communities composed mostly of cyber beings. These will have few human amenities.)

Many of these people will be railing against "The Man" (whatever he is called in the 2050's) and this feeling will support obnoxious social behavior. One 2010's example is using public restrooms as long-stay break rooms. What forms this will take in the 2050's is something these people will be diligently researching.

Commune groups

Most people will be living in urban settings, and many of these urban settings will be high-rise apartments. This is something I experienced in prosperous settings in Korea in the 2000's. What people will be doing in these settings will be quite diverse. But even with all the diversity being supported there will be some people who choose to be somewhere else.

Various groups with various causes will choose to forego these dense-pack urban settings and choose to live in distant rural settings instead. The reasons to get away from the crowd will be diverse, one common one is likely to be breakaway religious causes. An enduring example in the 2010's is the Amish people. These people will be practicing alternative lifestyles of various sorts.

The size of these groups will be small compared to the size of groups that stay in the urban settings, and they will not be as prosperous.

Space colonies

Space colonies are going to be the iconic distinct class of the 2050's. In the 2010's there are lots of people wishing and hoping they could travel to and live in space colonies. In the 2050's some of these are going to materialize. Some small ones will be science research-based colonies, but because cyber will be much better adapted to doing research in the various space environments, the human-filled ones will be few and far between. The most populous space colonies will be founded to support space tourism. They will be located on the Moon and feel much like high-class resorts do on Earth.

Conclusion

The technologies of the 2050's are going to be quite different from the technologies of the 2010's. This means that how people live and what they think about are also going to be quite different. In particular what people take for granted is going to be very different.

The result is that the communities of the 2050's are going to feel very different from those of the 2010's. The way people organize, their classes, are going to be very different. Some of these differences are foreseeable, but there are going to be many surprises as well.

Further Reading

- Three Kinds of Money by Roger Bourke White Jr., March 2017 - http://www.whiteworld.com/technoland/Visions%20series/2051-WIP/76-money.html
- Top 40 Jobs by Roger Bourke White Jr., December 2014 - http://www.whiteworld.com/technoland/stories-techno/stories2015/2015-TV3/2014-top-40-jobs.html

Immigration in TES

Introduction

Immigration is going to be dramatically different in the 2050's. One big difference is there is going to be a lot less of it. This will happen because the reasons to immigrate are going to be a lot less compelling when the Total Entitlement State (TES) is widespread around the world.

How immigration is going to be different is the topic of this essay.

Why people migrate

People migrate to better their lives. Migration is a tough process. A person is leaving familiar territory and friends to go to a strange land filled with strange people. It is often a frustrating, expensive and hazardous process. In spite of this millions of people in the 2010's undertake this every year.

Why?

Again, people are doing this to better their lives. They are seeking things such as better employment, reliable food, and a safer community.

In the 2010's the better employment is a critical feature. This usually starts with taking a job being offered by another person or company in this new and strange community. This job can continue for years, or as the person learns the ropes they can get even more ambitious, move up in employment, or start their own enterprises. The major benefit of starting their own enterprise is doing things even more their own way -- being their own boss. This is why immigration brings a lot of innovation to those communities which embrace it.

But in the 2050's things will be different.

Why the 2050's will be different

The biggest difference the 2050's will bring is the Total Entitlement State. When a TES is in place a person is entitled to food, shelter, health care and dignity just by being in existence in the community. The person doesn't have to work for a living, and fewer and fewer people will be working for a living as automation becomes more and more pervasive. People will be busy, but not busy at working on a productive job of some sort.

Achieving this kind of paradise has been an aspiration for many people for thousands of years. The big difference that will be coming in the 2050's is that, thanks to pervasive automation, it will be achievable and stable. And, again because of pervasive automation, it won't be people working that makes this possible, it will be automation, robots and artificial intelligence that are doing the heavy lifting that produces the goods, services and transportation that makes this paradise possible.

And this affects migration. What this means is that human jobs that are part of enhancing a community's productivity are going to diminish dramatically in numbers, and that means is that moving to get a better job is going to decline dramatically in magnitude.

This prime reason to migrate is going to disappear. Moving to get a better life is going to be an option for fewer and fewer people as time goes on.

Diminish but not gone

Economics is a major push behind migrating, but not the only one. Another one is the instinct to explore, which is a powerful instinct in many people. As TES builds everyone's prosperity the ability to explore just for exploring's sake becomes affordable to more and more people. This kind of migrating won't be as expensive, frustrating or hazardous as 2010's migrating for economic reasons is, so it will be conducted in very different ways and for very different goals. A common form will be taking school classes in a different land to extend one's education.

Conclusion

When people migrate in the 2050's, their motivation is not going to be the aspiration to get a better life by getting a better job. It is going to be something else, something completely different. This means that migration in the 2050's is likely to decline dramatically in magnitude and be very different in composition when compared to the 2010's.

And this means that innovation in TES communities is likely to be a lot less dramatic than what we are experiencing in the 2010's in those communities that embrace immigration. Human TES communities are likely to get a lot more Rust Beltish in their cultures -- complacent, but comfortable.

The "paycheck to paycheck" Lifestyle in TES

Introduction

The "I'm living from paycheck to paycheck." lifestyle is common in the US in the 2010's. It has been a common lifestyle choice for people for centuries. It is going to be even more common in the TES communities of the 2050's because getting paid is going to become even more disconnected from working in the traditional meaning of work.

The question in the TES environments of the 2050's becomes: What is this "next paycheck" these people are living for? What is the paycheck to paycheck lifestyle going to be like?

This is the topic of this essay.

Next paycheck living

Ever since there have been paychecks there have been people who live from paycheck to paycheck. This style is particularly common in young adults who like to party hearty when the paycheck arrives, but many others indulge as well.

As a community offers its members more and more social safety nets this lifestyle gets even more common. It is supported by both the instinctive thinking to celebrate good times when they happen and all the advertising a person encounters which is encouraging them to spend on luxuries of all sorts. That plus family nagging to get a "real life" becomes less and less meaningful. The converse lifestyle -- saving a lot for future-oriented activities such as house buying, expensive luxuries, more education and child raising -- is a learned skill, not an instinctive one. It used to be widely taught as home economics classes (now called family and consumer sciences classes), but these have fallen out of style and are now optional, not mandatory classes.

So paycheck-to-paycheck living has become common, and, as I pointed out above, as social safety nets get more pervasive this living style gets easier and easier to sustain. And TES lifestyles are all about social safety nets.

So... what is it going to be like in a full and stable TES environment of the 2050's sort? (There will be many different TES environments in the 2050's, not just a single monolithic one.)

Three kinds of money in TES

As pointed out in previous essays, I envision the TES world to have three styles of money, and these will be difficult to interchange.

The first style is Necessity Money. This pays for what the safety nets are providing. (One of the 2010's equivalents is SNAP, the Supplemental Nutrition Assistance Program -- what used to be called Food Stamps.) Necessity money is handed out to every community member in some fashion, but there are likely to be many different ways it gets handed out. (see more in the next section) There will also be lots of constraints on what can be bought with this style of money -- this is money with lots of strings attached.

The second style is Luxury Money. This pays for what ambitious people who earn money want to buy. This is paid out to people who still have a job in the 2010's sense of that term. And they can use it to buy things with a lot fewer strings attached than those that come with necessity money.

The third style is Investing Money. This pays for maintaining and building the manufacturing and service infrastructure of the community. As an example this pays for the conventional forms of streets, farms and factories we are familiar with in the 2010's. (Conversely, artisanal forms of farms and factories will get paid for in different ways such as using luxury money to support a hobby or small business.) I envision this investing money being controlled mostly by the cyber of the day. These cyber are the entities which are making the choices in how to run, maintain and grow the manufacturing and service infrastructures of the various communities of the world. Investing money will not be used as a savings tool by humans -- it will not be the equivalent of today's financial and stock markets. It will be a further evolution of the exotic finance forms being created in the 2010's such as unicorn financing.

Paycheck to paycheck is going to be mostly about how necessity money is spent.

Taxes in TES

Taxing will be mostly invisible. This is because so much of the financing that produces necessity money will come from cyber-dominated services and industries -- investment money domain. Most of the taxes levied in the luxury money domain will be of the VAT sort, and mostly low-profile -- they just happen, they aren't part of a human's spending strategies.

The exceptions -- the taxes that do stay visible -- will be those that sustain rituals. Ironically, one of those is income tax in the US in the 2010's. Tax preparation is an annual ritual activity -- exemptions and refunds stroke the "system gaming" and "something for nothing" instinctive pleasures. Because it is so ritualized it may endure in some fashion clear into the 2050's. If it does then the "Tattoos and T-Shirts" phenomenon says it will be even more ritualized in its 2050's incarnation. It will become part of personal expression.

Paycheck to paycheck in TES

Given the TES environment, how will necessity money be doled out? This is the kind of money that will be used most commonly for paycheck-to-paycheck living styles.

Some will be just handed out into accounts that can be used to purchase various kinds of necessities such as food and shelter. To the recipients it will just magically appear and then disappear again as things such as rent are automatically paid for. Necessary medical care will also be covered by necessity money, and the routine parts will be treated like food and shelter. The surprise parts -- accidents and unexpected ailments -- will be handled by a different routine, but also be part of the necessity money environment.

Much of necessity money will come with nagging in various incarnations. Along with the money will come constant advice on what activities the recipient should be engaging in. In this it will be much like Medicare in the 2010's. As an example, a recipient decides it is time for lunch. He or she goes to a restaurant and looks up and down the menu. As they are looking they get a reminder on their personal communicator that (for good health) they should either order a small portion or go out and do some jogging first and then come in and order a bigger portion.

Beyond the bare minimums money, a large part of the necessity money will come when a person goes through hoop-jumping to get it -- "You want [X] money, fill out this form and talk to this person."-type activities. This hoop-jumping will be a style of work in this TES environment.

Related to hoop-jumping will be system gaming -- activities which act like hoop-jumping but are supposed to be "sneaky" in some fashion, not routine. The cyber handing out the necessity money will see them as a form of hoop-jumping but the humans doing the activities will see them as gaining them something-for-nothing-style rewards. This will bring deep satisfaction to the humans engaging in the activities. It will also bring upon them the risk of social shaming if they are discovered by the wrong other humans to be doing these activities -- the 2010's equivalent is being revealed to be a "welfare queen".

Necessity food, shelter, and self-abusive practices

What are necessity food and shelter going to be like? What self-destructive habits will TES young adult folk engage in? (As in, what will be the "sex and drugs and rock and roll" lifestyles of 2050's?)

Necessity food

Because of widespread automation, necessity food providing is going to be centered around fast food style restaurant food choices rather than those centering on home preparation. Home preparation will become something artisanal rather than necessity. A person will go into a restaurant and look at the menu, or look at the menu on their internal communicator, and order. While they are looking over the menu their internal communicator will be informing them about which choices are available to them because they still have enough "paycheck" to afford them, and be advising them as to which choices are good for them based on their current state in their daily health regimen. The person makes their choice, chows down in the restaurant or has it delivered somewhere, and goes on with their life.

Personally preparing food will transform into an artisanal activity. Purchasing what is needed will be paid for with luxury money rather than necessity money. I say this with a "yes, but...". Because food is such a core topic of instinctive thinking, how it is treated will have a lot of emotion wrapped in it as well as rationality. Because of this some food preparation that should be artisanal will be treated as necessity "to help the poor", or as "a right" that necessity money should be covering.

That said, much of necessity food is going to be the burgers and fries of the 2050's discretely modified to be healthy eating for the basement dwelling gamer types who are the core of the TES lifestyle. Luxury money will buy whatever "organic food" is called in the 2050's. It too will be discretely modified to be healthy for whoever is consuming it. With both necessity and luxury foods, what is on the label is not likely to fully describe what is in the package.

Necessity shelter

The trend towards adult children living with their parents will continue and strengthen -- multi-generational family living will become more widespread. Likewise the trend for people to gather into just a few very large urban areas will continue. As a result in 2050 there will be:

- a handful of large mega-cities which hold most of the human population -- Shanghai, Seoul, New York City and a dozen or so more "where it's at" places
- among these will be many large-to-small cyber-inhabited cities -- these will be mining, manufacturing and farming company towns
- some resort-oriented cities -- some will be necessity-oriented such as Disneyland, some will be luxury-oriented such as Macao and Hilton Head

- places where small groups of people are living various alternative culture lifestyles -- the equivalents of 1960's Hippie communes
- Neolithic Parks

Neolithic Parks are places where the inhabitants are deliberately living Stone Age lifestyles and are deliberately isolated from other humanity. This is an insurance policy. They are doing so so that if a worldwide disaster destroys civilization as we know it there will be some people who can still survive in primordial conditions, and who can then thrive, spread and rebuild humanity.

In the mega-cities there will also be gentrified neighborhoods where young adults cluster together to have good times with their own age groups rather than living with parents.

Also mixed in with these other groups will be those who choose to live as homeless people. These homeless types won't go away because social justice warrior instinctive thinking will pay for people to keep indulging in the homeless/begging lifestyles. Begging will be one of the professions that survives and thrives in the 2050's. These people will be called the 2050's equivalent of homeless, but they won't be without adequate food and shelter -- TES will be providing for them. But looking homeless will be their version of cosplay. And they will be paid to do it in the same way homeless with signs on street corners are paid in the 2010's.

Necessity substance abuse

The instinct to party hearty until it makes one sick is a strong one in late teenage- and young adulthood. It's not going away. However, wearable technology is going to dramatically change how the hearty partying is done, and how much both short-term and long-term damage happens as it is being indulged in. This is going to be a big revolution in the lifestyle. Party hearty is not going away but it's going to be very different.

Because the harmful effects are diminished it will become one of the centers of TES lifestyles. The partying can go on long and strong for days... weeks... months... years even. This will become a core TES lifestyle.

And along with it another, "Yes, but...". The instinct to socially shame those who are indulging is also strong and won't go away. This conflict between indulgers and shamers will be similar to the War on Drugs conflict of the 1960's through 2010's -- the shamers will be saying "This is dangerous, don't do it." and the indulgers will be saying, "What danger? The danger is miniscule. And this is so much fun."

In addition to the lowered risk of indulging doing health damage because of the wearables, there is another lowered risk factor as well: The lowered risk of human lifestyle in general. Humans don't have to routinely do risky activities such as driving or operating other machinery. They don't have to drive home from a bar or wake up on a Monday morning thinking, "Oh God! I have to get to work."

But the activities risks are not gone entirely. Those who engage in physically-oriented artisanal activities as their work choice -- greenhouse farming -- and vigorously physical dilettante activities as their pleasure choice -- gymnastics -- can be engaging in risky activities. Those engaging in entertainment performances are an example of activities which often involve risk and "show must go on" feelings. These kinds of risks will still be around.

In sum, self-abuse is going to remain an activity which many humans spend a lot of time at in the 2050's. It won't be as dangerous as it is in the 2010's. But even with the reduced danger, the moralizing and social shaming associated with it will stay strong.

Advertising for necessity money purchases

Advertising for goods and services bought with luxury money is going to be a straightforward evolution from the 2010's practices. Advertising for goods and services purchased with necessity money is going to be something different.

First, let me define what I see is the goal of advertising: The goal of advertising -- when expressed from an optimistic point of view -- is to inform potential customers that a product or service exists, help them decide if it matches their needs and desires, and finally help them find a place to purchase it.

The different challenge connected with the necessity money world is: Who is paying for the advertising, and related, who is creating it and deciding where it gets placed? Compared to what is happening in the luxury money realm this is going to be strange and different.

It may be so strange and different that humans stay involved -- this may be an aspect of TES that cyber can't handle well. If that is so, what will happen is that cyber will take on the first steps in designing the goods and services they provide, and then come to humans to do the pre-introduction tweaking of the designs and handle the promotion of what is created. (once the product is on the shelf, cyber should be able to handle customer feedback as well as humans can) Advertising, like science and engineering research, may be an arena where there is lots of human-cyber interaction in doing the research and innovating activities.

Conclusion

How people get paid in the TES environment is going to be very different. How they spend is going to be quite different as well. One of the big differences is that necessity money spending is going to be accompanied by constant nagging as the money is handed out. The recipients are going to be constantly reminded of how they can be better citizens as they are forking out their unearned cash.

The issue of how a human living paycheck-to-paycheck makes choices in their necessity money spending is going to be an interesting one. How will they find out what they can spend on? (advertising) How will they know what they have left to spend at any given moment, and when is the spigot going to open again? (some kind of advice given by their cyber muse) How much additional benefit can they gain from hoop-jumping and gaming the system?

These are all interesting questions that each form of TES will have to answer.

The World of TES Living is one that still has a lot of research to be done.

The Shopping Experience in TES

Introduction

What is the shopping experience going to be like in the 2050's?

Thanks to Big Data and cyber, organizations offering products and services will be able to dramatically fine-tune what they offer to what each particular customer wants. Advertising will be more and more relevant -- good for both the consumer and those paying for the advertising.

But, what does this improved targeting get the organizations offering products, and what does it get the customers? Who, or what, is going to moderate what the product offerers offer so that it harmonizes with what the target customer needs, wants, and can afford?

A new skill comes into existence: harmonizing advertising and the target customer's budget.

Avoiding "Shop 'til you Drop"

This harmonizing becomes important because many people have a strong instinct to windfall spend, as in, "If you've got it, spend it! DRINKS ARE ON ME!" And in the world of TES lifestyles there isn't much training to curb this instinct. So if people aren't learning self-restraint on their spending styles, who will be doing their budget harmonizing? Who will be doing the restraining?

It is likely the customer's cyber muse will be helping with the harmonizing. It will be the muses who are teaching home economics-style good sense spending to people of all classes. This challenge applies to all classes and to both necessity and luxury offerings because the instinct to windfall spend is strong and pervasive.

Offering New Products

How does a new product or service break into the marketplace? How does Big Data understand and help new and innovative products succeed? This challenge applies most strongly to wildly innovative products that don't have familiar antecedents already out in the current marketplace. Examples of this issue: driverless cars have been widely predicted, but voting in Trump as a president in 2016 came as a complete surprise. How will Big Data handle Trump-like surprises in the 2050's?

Conclusion

Shopping in the 2050's is going to be dramatically different from shopping in the 2010's. "Smart" is going to make big changes to both what customers get offered in visible ways, and how they learn their home economics so they can decide what is within their budget and what they must forego for now.

There will still be surprises. In particular, how well innovative new products get accepted will still be an activity filled with uncertainty. And how these new products will mesh with smart advertising will be a constant surprise of the 2050's.

Charity and TES

Introduction

Supporting charity is sustained by instinctive thinking. This means -- even with the irony -- that it will grow in magnitude as TES communities grow in sustainability and prosperity. TES people are going to give to charities even though the TES state is providing all the basic necessities to all the people. The interesting challenge is forecasting what the charities of the 2050's are going to be like: how will they be promoted, who will do the promoting, and who will the donators think they are "saving" with the money they donate?

Top Forty charities

I envision a lot of "Top Forty" thinking going on here, which means the charity themes aren't going to change much from the 2010's even though the recipients' conditions do. There will still be lots of charities collecting money to give to poor children who live in distant lands, even though all those kids have their own smartphones. And there will still be beggars on street corners holding cardboard signs -- unless the driverless cars don't stop for them and there is no automated way to give handouts.

Again, the important thing to keep in mind is that charities will capture a lot of community attention and discretionary resources. Like entertaining and sports it will be something that captures a lot of dilettante attention and activity. This is going to be a dilettante activity that stays in the human community hands.

Politicians in TES

Introduction

Like all other human activities, politics is going to undergo big changes as we get into the 2050's. The biggest changes are going to be in what activities stay in the realms that politics influences. Because of cyber controlling much of big business activities, and because of cyber muses offering so much personally-oriented good advice, many activities that are currently influenced by the humans that are part of political and bureaucrat systems are going to move out of their control.

What is left for human politicians and bureaucrats to influence is the topic of this essay.

Politicians and money

How much will politicians and human governments control the spending of money? Most of the investing money, the big business money, will be controlled by cyber and be invisible to most humans, including politicians. How much will be left in the hands of governments and the politicians and bureaucrats that run them? What kinds of projects will they engage in?

An example of something that will be changing is taxes. Taxes can become pretty transparent, if people want them to be. Think of a VAT (Value Added Tax) assessed without mention in the price listed. The obstacle to this simplicity is instinctive thinking as highlighted by the truism, "The only two certain things in life are death and taxes." People expect to see taxes in various forms. In addition, taxes can provide a rich field for the system gaming many people love to indulge in.

It is likely that "major" taxing will become invisible and it will be handled with the "investing money" which is controlled by cyber. But humans will still support various "cosmetic" taxes which are in formats they can see, complain about and system game. These will be paid with luxury money. They are likely to be similar to the taxes-plus-deductions we have in the 2010's.

The instinct to complain about "greedy people taking advantage" is not going to go away -- it feels too good. The question becomes who are the greedy people? And how can they game the system in ways that allow lots of complaining, but little witch hunting?

And related, when stress and witch hunting time does come, how will the witch hunting be conducted? Who will be hunted and what will their shaming and punishment be?

Spending tax money

So there will still be taxes collected in human-visible ways. The next question is: What to spend those taxes on?

In the 2000's in declining cities of the American Midwest lots of spending was on high visibility projects like sports arenas and convention centers. Will those kinds of projects still be the center in 2050? Perhaps, because the visible taxes won't need to be spent on infrastructure upkeep -- that will be handled transparently by cyber and investing money. And, likewise, the taxes that in the 2010's are devoted to health care and pensions will become the necessity money of the 2050's. This money covers everyone including the poor, old and sick.

An example of the kind of high-profile ritual that visible taxes are likely to continue to support is bidding for, and investing in infrastructure for, things like Olympics and World Cups.

The Dark Side

On the dark side, governments, just as much as all other human organizations, are tempted to get on the spending Slippery Slope. In the 2010's Venezuela is the highest profile example of this happening. Will governments still be allowed to get on the slope in the 2050's or will cyber intervene in some fashion?

Losing the purse

How will politicians and bureaucrats lose access to the purse?

Will they be "taken off their allowance" when they go through an episode of not paying their bills, such as Puerto Rico is going through in the 2010's? Or will the loss come about in a lower profile and more orderly fashion? Losing access to the purse is going to be hard for these inherently proud people that politicians are. What will replace it in their pride spectrum?

And, looking at the even bigger picture, what will all these government roles still be needed for in the community? Will it be mostly for social shaming? Will it be for social entertainment? Related: As they lose access to the purse, what Big Vision-style projects can politicians promote? How can they, say, bid for an Olympic games? (as mentioned above)

Politicians and certifying

As cyber takes over big businesses humans will be doing more dilettante activities. Government can stay involved in dilettante activities by certifying them.

In the 2010's this is an activity that local governments get involved in. The highest profile is issuing driver's licenses and auto license plates, but all sorts of other activities get licensed as well.

And there is much complaining by libertarian types that governments are way too involved in this activity. The commonly mentioned example is how hard it is to get a hairdresser's license in some localities such as cities in California.

The 17 Feb 18 Economist article, Occupational licensing blunts competition and boosts inequality, talks about how occupation licensing is booming in the US.

From the article, "Occupational licensing—the practice of regulating who can do what jobs—has been on the rise for decades. In 1950 one in 20 employed Americans required a licence to work. By 2017 that had risen to more than one in five. The trend partly reflects an economic shift towards service industries, in which licences are more common. But it has also been driven by a growing number of professions successfully lobbying state governments to make it harder to enter their industries. Most studies find that licensing requirements raise wages in a profession by around 10%, probably by making it harder for competitors to set up shop."

Licensing dilettante activities that don't have much effect on cyber-controlled big business activities is likely to continue. This is something humans can be doing that won't affect core prosperity, but it gives a lot of sense of accomplishment to those issuing and those receiving.

Being the social conscience

One of the activities politicians routinely engage in is expressing elements of the wide range of ideas that regularly pop up in community thinking. When the ideas are crazy enough the media makes small news by mocking them. But expressing these wide ranging ideas is part of the function of politics, in particular the legislative branches. This is very human, and it will continue.

Conclusion

The role of politicians and bureaucrats is going to change steadily and dramatically as we progress into the 2050's. As cyber takes over more of the mainstream productivity functions these will move out of the hands of politicians and bureaucrats because these will become low profile activities in the human awareness spectrum, and because they won't understand what is happening well enough to intervene meaningfully.

Instead these people will devote their attentions to what remains high-profile in human awareness. They will devote themselves to intervening in dilettante and Top Forty human activities. Their function will be to license, promote, shame, and provide a forum for various proposals to be aired that are inspired by social conscience thinking.

Further Reading

- The Temptations of The Slippery Slope by Roger Bourke White Jr., July 2015 - http://www.whiteworld.com/cyreenikland/editorials/editorials2015/2015-rent-seeking.html
- Occupational licensing blunts competition and boosts inequality, Feb 17th 2018 - https://www.economist.com/united-states/2018/02/17/occupational-licensing-blunts-competition-and-boosts-inequality

Social Shaming in TES

Introduction

Social shaming is an activity that has been around as long as humanity. It was as routine in the Stone Age as it is in the 2010's. But the forms and functions it takes vary with each society and each age. Technology and enfranchisement both affect the styles and influence of social shaming.

This essay is about the likely styles and influences of social shaming in the communities of the 2050's -- the necessity and ambitious class lifestyles.

Background

Social shaming is a tool a community has for influencing the behavior of community members. It is a negative tool, encouragement is a positive tool.

Social shaming is context sensitive -- consider the difference between what people shame at a Sunday church meeting and at a raucous party-hearty victory celebration for the home team winning a championship.

Social shaming is influenced by community-wide feelings such as enfranchisement, complacency and embracing a Big Vision. Enfranchisement and embracing a Big Vision make people of the community more tolerant while increasing complacency makes people more prescriptive. An example of Big Vision creating a tolerant environment is the boom-town environment. In a boom town a Big Vision is in place and active -- the people of the community are there to get a big project accomplished. This makes them more tolerant and less prescriptive. An example of this boom town tolerance that comes to mind is Margaret Bourke-White reporting in Life magazine's first issue (1936) on life in the boom town that was building the Fort Peck Dam in Montana. One of the pictures shows a child sitting on the counter in a bar. In communities not participating in a Big Vision this is a shamable activity.

And social shaming is technology sensitive. In the 2010's we have Facebook and smartphones as well as over-the-fence gossip for conducting shaming campaigns.

As we roll into the 2050's our lifestyles will be changing in many dramatic ways and one of the changes will be in how we conduct our social shaming -- what we shame and how we shame will both be changing.

What's coming

Social shaming will be a big part of 2050's life, in particular necessity class life. Shaming is highly emotional so, like love, engaging in it is comfortable for lots of people. Historically what held it in check was activities that were dangerous and required lots of serious cooperation, such as crewing a sailing ship or working in a factory. But if you and your associates are not working on anything dangerous and requiring serious cooperation, why not let it get big?

A necessity class lifestyle is specifically designed to be not dangerous and not requiring lots of human cooperation to be prosperous. This is an environment where all the basic necessities are being provided whether or not a person cooperates with others.

The ambitious class lifestyle is the converse. This is one where a person is working on something critical to improving prosperity. In this lifestyle enfranchisement and a Big Vision will be operating and tolerance will be much higher.

In the necessity lifestyle, where you don't have a Big Vision in your life to keep you cooperating with folk around you, then acrimony ("trolling") can suit just fine for day-to-day living. Acrimony is supported by the Us versus Them instinct.

And this brings up the question of how is busybody prescriptivism going to fit in? Who gets to put their nose in other people's business, and how do they do it? How forceful do the busybodies get to be?

This is likely to be the biggest threat people feel in the 2050's. This will be mixed in thoroughly with pervasive surveillance. Related: How will cyber muses get involved in social shaming? Which way? Will they diminish or amplify?

Prosperity in the 2050's won't depend on lots of humans cooperating to accomplish a Big Vision, so the chances of one emerging in the necessity communities are pretty slim. The best chance of a Big Vision happening is that cyber, for some reason, sees benefit and creates and supports a Big Vision for the necessity humans to follow. This could happen because people following a Big Vision are in general happier than those being acrimonious, and cyber muses are working to keep people happy.

An example of a region and time that appears to have been good at sustaining Big Visions is the Nile River region during its Ancient Egyptian cultures, with their monument building being the enduring product of their stream of Big Visions.

Shaming: The Brothel Model

Brothels and the sex trade are a historic example of social shaming where the participants and what they do are known to the community and the practice continues in a stable fashion. These people are shamed but they go on with their activities in spite of the shaming. This is what stable 2050's social shaming will be like.

I foresee that sustaining the sex trade in an environment of pervasive surveillance will likely put it in a pioneering role in how privacy will evolve as the pervasive surveillance environment becomes more and more widespread. Who gets to see what is going on in the brothel? Who defines what is a brothel? Who will tolerate what they see? Who will shame what they see? What will be formal shaming, what will be informal shaming?

In sum, shaming in the necessity communities is likely to be active and widespread. It will be "trolling" in those many guises that are active in the 2050's.

Wearables and shaming

How will wearables influence shaming? In the 2050's wearables will be able to control emotions -- a person will be able to "dial up" or "dial down" various emotions such as love and fear. Shaming is highly emotional on both sides, initiating and receiving, so wearables should make a big difference. But what effect will a wearable that is controlling those emotions have? I'm guessing it will be able to make the receiver thick-skinned. Can it also reduce the joy a troller gets from initiating insults? What will make the troller want to do the reducing?

Conclusion

Because the danger and the benefits of people cooperating are going to be reduced in the necessity lifestyle, social shaming will increase. It will take on a much bigger role in many people's lifestyles than it does in the 2010's.

Thanks to lots of new technologies how to shame will become more diverse and what to shame -- what to get prescriptionist about -- will also increase in diversity. People are going to be watching their steps a lot more closely in the 2050's -- their steps and the steps of those around them.

The Stories

The Unexpected Hero

Jimmy John is in his basement immersed in his VR gaming.

He is in "Fall-in 14: Hometown" a first-person shooter game. He is in the post-apocalypse ruins of West Valley City, his home town and where he currently lives.

And he is in a tight spot. The zombies and creepies have run him into the ruins of the Starscents across from Valley Fair Mall. He is on the second floor, wounded, out of healers, low on water and ammo. If they find him like this, and swarm, it is Game Over, and start again from his last resurrection at the Valley Hospital. He will lose all the progress he has made in the last two hours of gaming, and he is not looking forward to that.

"Grrr." ...but what can he do?

Then from off screen comes the sound of automatic weapon fire.

"Good! That isn't zombies." he thinks.

The fire gets closer, he risks a peek out the window. Someone in power armor with a hot automatic weapon is running towards the Starscents, gunning down all in his way. He goes in, below Jimmy, and shoots the place up, then comes up the stairs.

They are now face-to-face.

This newcomer pulls off his helmet. The face is handsome. He smiles at Jimmy.

"Hi. I'm Chris Romney and I'm running for mayor, in the real world. I hope I was helpful. Remember to vote next month. Now I'll be on my way. I've got a few more constituents to rescue before I call it day."

"Nice armor!" says Jimmy, looking him up and down.

Chris grins back, he really likes hearing that. It means he will be remembered, and there is just a bit more twist to this, "Remember Me" Plan of his…

He says, "Yeah! Hot stuff. I'm lovin' it. But I'm just renting it from Armors R Us over on 3rd West and 2700 South. I'll be taking it back to them when I leave this world."

Chris puts his helmet back on, waves, bounds down the stairs and heads out.

Jimmy, much relieved, follows him out and then heads for his relief camp to heal up and restock, he will now get there without difficulty. And, with that armor for rent in mind, he now has a new adventure he can look forward to.

Game saved. Impression made.

Jeanie The Gene Editor

Introduction

Jeanie thought of herself as an anti-nerd.

She didn't spend hours on mastering Commander Codeine v15, instead she worked hard to understand the real world. She was born and raised in a free range kids style baby club so spending time outdoors in the real world was easy to do. As a young girl she spent a lot of time messing in the rocks, mud and reeds of the stream that ran through a wild park beside their apartment complex. While she was really young she mastered the art of herding tadpoles in that stream. Then as she grew older she saw that they sprouted legs and lost their tails to transform into the frogs she also saw. As she grew still older she switched from tadpole herder to frog hunter, loved that, and then moved on to other interesting outdoor activities in other places, like exploring forests, cliffs and waterfalls in the parks her family visited, and getting into swimming and sand castle building on the beaches. She loved exploring coral reefs.

Now as a teenager she was discovering neat things and finding adventure indoors as well as outside. She liked what she was learning in school, and one of the things she discovered there was gene editing. She further discovered she could feel a lot of accomplishment making those twisty DNA and RNA molecules do what she wanted. And when the course at school ended she discovered she could do this at home, too. She could buy do it yourself (DIY) gene editing kits. Well, her parents actually bought them, they were pretty expensive.

Using them she edited genes in E. coli and other bacteria so they would make new kinds of proteins. She edited viruses so that when they infected the bacteria they would do the same -- for some proteins that was the easier method.

Her current foray is gene editing bacteria that will then live inside a worm's intestines -- gene editing gut bacteria. The goal is to change the worm's health by changing what proteins the gut bacteria make and excrete. Some of the new protein will get absorbed by the worm's intestines and become part of the worm's body chemistry. There it will affect the body's metabolism and through that the body's health and well being.

This is an indirect way of manipulating a worm's health, but it is much simpler than trying to directly manipulate worm DNA and genes. The bacteria and virus DNA and genes are much, much simpler in their layouts. She aspires to work on mammals someday but they are really expensive and lots of people still worry that there is a gene apocalypse just around the corner. They worry that DIY editing will create some unstoppable evil. (and not just DIY, by the way) But with the cost of gene editing equipment dropping so steadily, and the precision rising so steadily, and so many styles of genes to work on that aren't connected to humans, those worries are drowned out by the ease of getting into the hobby.

Net result: Jeanie is having fun exploring a new technology, and doing yet more learning about the real world.

Wise Old Man -- JC version

Introduction

Every awakening is such a strange event.

But some are stranger than others, and this one is a doozy.

I'm coming out in the body of a young boy dressed in primitive clothing. I look around and the technology around me is... medieval? Wait! Pre-medieval? It seems to be! I'm dodging puddles as I run down an alley made of dirt... just dirt! The buildings to the left and right of me are made of rough-hewn wood on the right and adobe clay bricks on the left, that's it. There is trash all over in the ally and not a lick of metal in it.

What is this world? And what will I do here? What wisdom will I be adding to this land solidly in the Agricultural Age?

This was going to be a new experience, indeed!

I watched. I watched for twenty years, while I learned a whole lot about how things got accomplished in an age without computers, steel or engines.

And with time the events of the day started sweeping me up. I could see that the politics of the day were calling for wisdom, wisdom of the sort I could provide... but at some personal risk. (Well, not that much. People here died left and right all the time from disease and accidents of all sorts. Violent death was just a part of the usual mix.) The land I was living in had just been conquered by a group of outsiders from a distant land across the sea that was more technologically advanced -- they had big ships, hefty armor and iron weapons. They also had a more organized social order. As

I watched what was taking place after their soldiers and governors moved in, I could see my people were becoming part of a big and rapidly growing empire. That empire was showing my people not just iron weapons, but how to do lots more trade with distant lands.

Lots of changes. There were new jobs because these imperialists wanted our people making things they could take to distant lands. Some people were making a living at these new jobs, a handful are getting filthy rich. A dozen or so were getting filthy rich doing a particular high-profile job that the rest of the community didn't think was a good one -- money lending.

People around me were grumbling because that dozen were making money trading money, not doing real work.

"It is an outrage!" they say to each other, and me. And this particular outrage comes on top of all the stress of the change brought by this imperializing.

For the last couple years I have started coming across as a "smart guy". I have been placid about adapting to this imperializing and been offering good advice to others in the neighborhood. I'm now a craftsman by trade, and that plus my good advice is getting me lots of local respect.

When I get asked to help out with this "rich trading man outrage" problem, I decide it is time for some wisdom. I try to talk with this rich dozen…

These guys are in a courtyard that is just outside the local central temple. This temple is a big one, and really fancy -- it is a marvel that shows just what can be accomplished with stone, clay and wood, and a lot of skilled craftsmen who are willing to devote years and years to a building project. Even with all I have experienced, I'm impressed.

Inside the courtyard it's a busy time for these money changers. (Well, they are always pretty busy.) I have to stand in line a long time. And when I get to the head of the line…

"OK! What kind of coin do you have?"

"I'm not here to trade. I'm here to talk. There--"

He gives me a big raspberry and shouts, "Pffft! Move on! Rube! You want to talk, go find your momma. There's a lot of people here who want to do business." He points to the person behind me, "Next!" He also motions to a guard who is now moving to escort me away.

He was not happy to hear me saying I wanted to just talk. I was not happy with him brushing me off in that insulting way. I topple his table.

"See how busy you can be now, Money Changer!" and I storm away.

The people who came with me, the angry ones, took my cue and toppled two more tables. Some of the customers hit the dirt and started grabbing for the coins that were scattered. It was a good thing I was so upset, I just walked away, angry as hell. I reached the archway and was outside counting to ten before I noticed the bedlam inside and the guards rushing by me to get in and restore order.

I chuckled a bit and moved on. This was now high profile enough that I had done all I needed to do here. I was wise enough to see that!

That weekend I gave a talk on a hill, I had started doing these weekend talks a couple months ago. My cyber muse had recommended it. With all that temple ruckus, I got a big audience this time.

There were a whole lot of people, which meant a whole lot of strangers, but a handful stood out. They didn't move like locals. I was getting some cyber support on this speech.

I talked platitudes, platitudes for me, anyway. But they were platitudes these people really enjoyed hearing. And mid-way through, my cyber associates started handing out bread and fish. All-in-all, it was another high-profile success.

Then the trouble started for me. The local authorities were not happy with all the recognition I was getting. I got hauled in front of the local religious leaders.

They wanted to hang me up by my thumbs. But when they saw the crowd outside who liked what I had been saying, and saw how "enthusiastic" they were getting, they decided to push the matter upstairs, to these new imperialists, and let them take the heat.

This new governor was not at all happy to see me in front of him. He too saw the crowd outside. And he openly wondered why this wasn't being handled by the locals. "He hasn't been bad-mouthing the emperor, why is he here?" he wondered.

But the crowd outside made him nervous enough that he decided a show of force was necessary. I was dragged to a local hilltop and hung on a post to show what happens to "troublemakers".

Ouch! No fun at all! Even less fun when one of his soldiers sticks me with a spear while I'm hanging there.

I die soon after, and soon after that I get pulled down by my cyber associates who take me into a nearby cave they are using as a base. Inside, they revive a copy body of mine, a full grown version, and transfer my consciousness to it. It takes three days.

...In truth, I don't remember that last part. The memory recording was turned off after the governor made his choice and I was being led out. The cyber saw no need to add what followed to my experience inventory! I saw some videos of it taken by the cyber, so I know what happened and what I looked like. I agree with their choice. That third-party viewpoint was sufficient for me to experience.

Once I'm fully in the replacement body I come out and show myself off briefly. Yeah, it's sure impressive for the locals to see that happen!

That showing up again is sure going to make my words memorable... but exactly how they get remembered is something only time will tell. There are going to be lots of arguments about what I said, and what I meant by what I said. That is something my wisdom lets me know in no uncertain terms.

With my brief reappearance my job here is done. I'll be remembered, I have imparted some wisdom, and I've given the locals a lot of new hope and meaning in their lives.

I 'git while the 'gitting is good, and leave my mark on local history.

All-in-all, a most curious experience for a wise man. And one of my more memorable, I have to admit.

Alien Girlfriend

Summary

A lighthearted story about an alien who comes to Earth to experience being a human falling in love. Deam Garetmar, the alien, captures Andrea Mathews, a high school girl, and makes a replica of her body which Deam will control to go flirt with Bobby Bantam, Andrea's boyfriend. But Ouch! Bobby has just dumped Andrea for a new girlfriend. Deam and Andrea now conspire to get the boyfriend back.

Chapter One

Deam Garetmar's flying saucer flits into the Solar System seemingly without a care in the world... well, solar system actually. It moves from place to place, admiring the clouds of Neptune, the crazy tilt of Uranus, and the wild, wild rings of Saturn. It moves deeper and deeper into the solar system. It is headed for Earth. It flits by the Hubble Space Telescope and looks at it with great curiosity and interest -- its camouflage is so good that the telescope can't return the favor.

It arrives at Earth, flies down, and lands in a thick forest in Dewar State Park, not far from Bayzleville, Ohio -- still undetected by anything human. Once landed Deam gets out of the saucer and breathes deeply of Earth forest air. She isn't actually breathing it, but her space suit does some analyzing on what comes in through the vents. She does this because she has seen humans do this in videos of human people ending long trips. Deam and her people have been listening and watching the radio and TV broadcasts that have been beaming away from Earth for over a century now. They are up to the 1950's in what they have seen.

Deam is an alien, but a rather human looking one. She... *it* actually... is bipedal, about five feet high, with big blue eyes and with gray shiny skin -- underneath her space suit, that is. The suit is form-fitting so her body shape is easy to see, and the helmet is all glass, or the alien equivalent, so her face and head are easy to see, too.

The breathing ritual taken care of, she begins preparations for her mission here on Earth. Deam is here to catch an Earthling and do some experimenting on it -- some experimenting that is near and dear to her heart. Deam hides next to a nearby trail that looks to be lightly traveled. She wants to catch just one human, not a crowd of them.

After just an hour of waiting, the perfect experimental subject comes jogging up the trail. This subject is Andrea Mathews, an attractive high school senior who turned eighteen a month ago, and in addition to looking good is likely to be class valedictorian. She is here on a take-a-break afternoon hike.

Deam zaps her with "stun" on her ray gun and catches her! Deam takes her back to the flying saucer, into the laboratory inside, plops her down on the examination table, and ties her securely to it. Phase One complete!

Andrea recovers shortly after Deam secures her to the table. She looks around and is amazed at what she sees.

"Where am I? What am I doing here?" she looks at Deam, "Who... What... are you?"

It takes a while to happen but Deam smiles back -- another trick she has learned from watching TV -- and now she is getting to attempt it in a real time response. It's tricky for her.

"Sit still now, dear, I'm going to make a replica of your body."

"A what? This is crazy." Andrea struggles but to no avail.

Deam keeps the smile, "Again: sit still. The more you move the longer it will take for the replica to be completed."

"Why are you making a replica of me?" Andrea asks.

"Because I want to experience being a girlfriend." Deam explains, "I'm an anthropologist on my world and I'm fascinated with your human love stories we have heard about in songs and seen on TV. I want to try out being in love."

For a while Andrea is speechless, "This is crazy." she finally says.

Deam practices her smiling again, "For you. But for me it is exciting."

"Look... over there." she points across to the far side of the laboratory. There an alien machine is busy. It has lots of robot arms and other devices. In the center it is glowing and a human body is taking shape. It looks identical to Andrea. "That is an avatar. It is remotely controlled. I will be controlling it to take your place. It's going to visit your boyfriend..." Deam looks hard at Andrea, she is studying her, and reading her mind, "Bobby... Bobby Bantam."

Andrea is amazed, "How did you know that?"

"I read your mind." Deam answers back and gives her another smile, "I will read enough that I can act just like you. And I can experience being in love." for a moment Deam thinks dreamily about that prospect, then snaps out of it and gets back to business, "And then I will write a report on it. It is my thesis topic."

"There isn't going to be much loving involved." says Andrea, "I'm a virgin. And you better keep me that way!" she threatens.

At first Deam is mystified. She does some more mind reading, "Oh! OH... So that's what all that roundaboutness has been about! That was sure one of the odder parts of the stories." She thinks a bit, "OK. I think I can work with that... It should make this more interesting, in fact."

A beep comes from the machine on the far side. Deam jumps a bit in delight. "The replica is complete." She goes to a glass-sided controller on another side of the laboratory, gets in, and shuts the door. Once inside she puts on some controllers, and in a minute the replica opens its eyes and sits up on the platform where it was created. The replica is a totally naked version of Andrea.

It shakes its head and moves its hands and arms in an awkward fashion. It says, "My! This is going to take some getting used to." This is Deam talking through the replica. It awkwardly stands up and slowly walks around the room, almost falling over a couple times. "Yes, this is definitely going to take some getting used to." The replica moves to a chair and manages to sit down in it. Then it goes limp, and Deam gets out of the controller. She walks over to Andrea.

"Let me get you into something more comfortable," she says. She releases Andrea from the table and moves her into a holding cell on the last wall of the laboratory. It looks much like an Old West jail cell. "You'll be more comfortable in here, and you can watch what's happening on that screen over there. Oh, and take all your clothing and stuff off. I need to dress the avatar."

Andrea considers for a moment, then strips -- there doesn't seem to be any reason or hope in objecting. Everything comes off and she hands it to Deam. Deam puts the pile of stuff in front of the replica and then gets back in the controller. She brings the replica to life, and, in a really clumsy way, the replica gets dressed.

"Whew! Lots to learn," says Deam as this dressing goes on. Andrea can't help but smile. This is now looking like some kind of comedy movie.

"I hope you are a fast learner," Andrea comments quietly.

Deam is, and in just five minutes the replica is dressed and walking around the lab. "Much better," says Deam, "I'm going to take it outside now and finish your hike. You can watch on the screen."

The replica walks out of the flying saucer and heads for the trail Andrea was on. The walk is clumsy, like a zombie, but it steadily gets better as each minute passes. Once on the trail it heads back to Andrea's car. When the car gets in sight on her screen Andrea gasps. There is fearfulness in her voice when she shouts over to Deam, "Wait! You're not going to try and drive that, are you! You can hardly walk. How are you going to learn how to drive! You'll wreck my car!"

The replica pauses as Deam thinks about this problem.

"Hmm... You're right. This is a bigger issue than just falling on my ass every so often." She thinks some more, then makes a choice. "OK... I'm going to let you run the replica when it is driving the car."

Andrea's eyes widen, "I can do that?"

"Just like I can," says Deam confidently, "but, I guess you'll have to do some learning, too."

Deam gets out of the replica controller, lets Andrea out of her cell, and helps her get into the replica controller. "Give it a try," she tells Andrea.

For about five minutes Andrea is as clumsy as Deam was, but this is a human body replica so she adapts much more quickly and completely. In ten minutes she is ready to get behind the wheel. She does and drives the car home. When she gets there, Deam takes over again.

"Now, lets meet up with your boyfriend," she says this great anticipation.

Chapter Two

Deam walks to Bobby's house. It is only a block away. As she does she gets a thrill of excitement coming from the replica.

"Oh my!" she says to Andrea, "These hormones that surge through your body when you think about Bobby are quite exciting and powerful. Now I'm understanding why you humans sing so many love songs."

As she walks she gets even more in the mood.

"Umm... yes, indeed! I am! I feel like singing myself now."

"Don't you dare try!" says Andrea, "I don't sing. And you know how clumsy you are."

Bobby's door comes close before Deam can break into song. Without hesitation she walks up to the front door. She has her hand on the door handle when Andrea shouts, "KNOCK FIRST! And then wait." Deam knocks and waits.

Bobby Bantam comes to the door and opens it. He looks quite surprised.

"Andrea," he says in a neutral way. He can't think of more to say. He distinctly does not invite her in.

As he is standing there another high school girl comes up behind him. It is Betty Lou Meyer. Andrea recognizes her, she is in the same class as Andrea and Bobby are. Betty looks very comfortable as she walks up behind Bobby, she puts her hand on his shoulder and gives Andrea replica a big "I win" smile.

As she does Bobby finally finds some more words, "Betty and I are... are doing some studying for the test tomorrow." and he adds a big hint, "It's not a class you are in."

Andrea gets it, but Deam doesn't have a clue. Andrea is surprised and outraged at this. Bobby is dumping her for Betty! Sure, she and Bobby had just started hitting it off only a month ago, but this was not on her "could happen" list! In addition to being a complete surprise, she can't think of any quick way to explain this to Deam, so she remains silent, fumes, and just watches this evolve.

Deam doesn't have a clue. She is feeling all the warm fuzzies the Andrea replica is feeling. She smiles a big lovey-dovey smile at Bobby and says, "What class is it? Maybe I can help." And being Deam, with all her alien intelligence, maybe she can.

"It's astronomy. You're an art major." Bobby says it judgmentally and then shuts the door on her.

"But... but, I love astronomy," says Andrea replica to the closed door.

"Let me get in!" shouts Andrea back in the lab. Deam sees that she is clearly not in touch with what's going on so she lets Andrea get back in the replica. When Andrea gets in, the replica now starts filling up with her new emotions -- the sadness, the outrage, the humiliation. Andrea walks the replica away from the door to a quiet spot on the sidewalk between two houses.

"OK. You can take over again," she says, and Deam does.

When Deam gets in Andrea replica first screams, then starts crying. "What! Wow!" says Deam. This emotion surge is sure a surprise to her. But then she recalls some of this in the TV dramas she watched. Recalls, but Andrea replica doesn't recover.

"Take me back to my room," advises Andrea. Deam does, with Andrea replica crying and sobbing all the way. She rushes into the house, up the stairs, and into her room. Andrea's mother is in the kitchen.

Chapter Three

"Put me on the bed," advises Andrea.

Deam does so, then pops out of the controller. Andrea replica is lying face down on the bed still sobbing, but just quietly now. Deam crosses the lab to the cage, looks at Andrea, and says, "What now? What's happened?"

Andrea says in a straightforward way, "Bobby has dumped me. Betty is his girlfriend now."

Deam thinks about this, then says, "What happens now?"

"Well... nothing. I get to go find a new boyfriend. That may take a while... a long while. The school year is ending, and next year I'll go to college. It will be somewhere completely different, with lots of different people."

"Ouch! This will kill my thesis!" Deam looks desperate,

"Is there anything we can do?" she thinks and says, "Can we get Bobby back in love with you? That could be quicker than finding someone new."

"Hah! Good luck with that!" snorts Andrea, but she follows this shortly with, "Well... if you... we can. I'd sure like it. I still think Bobby is super. Well... he can be super... if he gets over this Betty business real, real quickly."

Just then comes a knock on Andrea's bedroom door back in Andrea's home. Deam dashes back to the controller, gets in, and starts controlling Andrea replica just as Andrea's mother opens the door and comes in.

"Is everything all right, darling?" she says as she comes over to the bed and strokes Andrea replica's hair.

Deam stops Andrea replica's sobbing, rolls over and looks up at Mom. She isn't sure what to say, but shortly something comes out, "Oh, I'm fine, Mom. I'm fine."

Mom strokes her hair some more, "What happened?"

"Oh, Bobby and I got into an argument. I came home. I'll be OK."

"Over what?"

Deam can't think of anything better to say so she blurts out the truth, "He wouldn't let me help him study for a test."

Mom thinks about this, "Sounds strange."

"It's an astronomy test."

"Ah... that makes more sense now." Mom decides this is just teenage angst, and it will pass. "OK, dear, dinner will be ready in an hour." She gives Andrea replica a kiss on the forehead and heads back to the kitchen.

Chapter Four

Andrea replica rolls over on her back, gets thoughtful rather than sobby, and says, "OK, what do we do now?" It is Deam doing the talking.

Andrea says, "We need more information. Can you go spy on Bobby and Betty?"

"The closest thing I have is this." she has Andrea replica slap her thighs.

"Well... let's go." says Andrea, "We can sneak over and look in a window."

Deam likes this plan. Quietly, so as not to disturb Mom, Andrea replica heads out the door, down the stairs, and out of the house. This time they head over through backyards, and when they get to Bobby's house Andrea replica climbs up a tree next to the house and sneaks a peek into Bobby's room.

There are textbooks on the table, but Bobby and Betty are not paying them any attention. They are giving wholehearted attention to each other, but with all their clothes on.

"Grr…" says Andrea quietly in her cage. She doesn't say any more, but she is wondering if she was controlling the replica if she could jump from the tree into the window and give these two a swat.

Deam is looking at this scene with a cooler head. "How are they going to pass the test?" she says.

"Clearly that's not the mind of either of them, right now. From what I'm seeing, they won't start thinking about that for at least an hour."

"Well, I can abduct Betty. Take her back to my world and experiment on her."

Andrea shudders at that, "Ouch! That only sort-of-solves half the problem. Yeah, Betty is out of the way, but Bobby will still want her. And especially so if he finds out she's been alien-abducted."

Both look and think some more. Bobby starts putting his hands on Betty's chest. She closes her eyes and starts breathing deeply.

"That hussy!" gasps Andrea, "What a slut!" after a pause, "I guess that's why Bobby likes her… Hmmm…" Andrea is thinking hard, "That could be what Bobby sees in her… And what she sees in Bobby… If so… Hmmm." the hard thinking is turning into a plan. "Maybe we can find a boy Betty likes more than Bobby. Now it's pretty clear what she likes." Andrea has another thought, "You better take me back to my room now. It's almost dinner time."

Deam climbs Andrea replica down from the tree and quickly sneaks back to Andrea's room.

Chapter Five

In the few minutes while Andrea replica is waiting for dinner in the bedroom, Deam comes over and studies Andrea in her cell. She is reading her mind some more.

She ponders out loud, "If Bobby likes Betty for touching her boobies... how about if we find a boy Betty likes even better who wants to touch her boobies? This could solve the problem this evening."

Andrea replies skeptically, "Just who do you have in mind?"

"Looking at your mind... I'd say Jimmy Galore would do the trick nicely."

Andrea gasps at that. In her mind Jimmy is the hottest item at school. How she longs for him when she sees him in the hallway.

"If you're going to set up Jimmy with someone, set him up with me!" she replies.

Deam smiles that, "That would be satisfying, wouldn't it... if it worked out long term... but that project would take too long. I need you in loving arms tonight, not two weeks from now."

"So how do you propose to get me back in Bobby's arms tonight?" Andrea is still skeptical.

"I have a plan." Deam's smile gets even broader.

Chapter Six

That night after dinner Andrea replica slips out again and heads for Bobby's house.

While Andrea replica is sneaking Deam tells Andrea, "I've been making some phone calls using the flying saucer connections. They have a lot more um... capability... than your phone does. Watch this."

At Bobby's house Andrea replica climbs the tree again and knocks on Bobby's window. Surprised, he comes and opens it.

Andrea replica says, "Come on out. There's something I want to show you. It's just down the block."

"What is it?" says Bobby suspiciously.

"Something you really, really want to see."

He looks at getting out on the tree with Andrea replica but decides to use the back door instead. He comes out, Andrea replica climbs down, and off they head to Betty's house.

There in the backyard, being just a little discrete, are Betty and Jimmy. Betty is in Jimmy's arms. Not just in them, she's melting in them, and Jimmy has his hands on her chest. As Bobby and Andrea replica watch, Jimmy's arms slip around to Betty's back while his lips are still on hers and he tips her over backwards in a dramatic Hollywood-style kiss.

"That hussy!" whispers Bobby.

Andrea replica gets in his face, puts her arms on his arms, and whispers, "Let's do them one better."

After just a moment of amazement, Bobby thinks this is a great idea. He pulls her to him, pulls her lips to him, and Andrea replica melts in his arms. She sighs, and back in the flying saucer Deam sighs. After a moment of watching all this sighing going on Andrea joins in.

Chapter Seven

It is late, late that same night. Andrea replica comes into the flying saucer. Deam releases Andrea from the cell. "Get dressed," she tells Andrea and Andrea replica strips down.

"We're all done now, Andrea," says Deam. She smiles once more. This time there is a gasp with the smile as she remembers the hot time she has had with Bobby. "You have quite a boyfriend there, at least in my opinion. And I now have a really hot thesis to bring back to my class." She strokes Andrea's hair, "Thank you, my dear."

Andrea has her clothes on now, "You think I can make it with Bobby?" she says.

Deam looks at her admiringly, "I think you two can make it really big time. My guess is: I may come to envy you."

Andrea is ready to leave. Before she goes she gives Deam a kiss and says, "I wish you good luck on your thesis. And I wish you good luck on whatever you have that is like me and Bobby being together and in love.

"This has been quite a day, all right."

Andrea yawns, it has been a long day, and she rushes out to head home. Test day tomorrow at school.

Once she is out of sight, the flying saucer flits back into space and heads home with Deam still smiling.

He's The One?

Preface -- the technology setting for this story

This is a story taking place in the America of the 2050's. This is a world filled with cyber muses and 3rd generation wearables. Cyber muses are AI designed specifically to interact with humans and inspire them. These wearables of the 2050's can both monitor a person's health in many ways and adjust it by adjusting health parameters such as heart rate and the levels of chemicals in the blood and organs, such as salts, insulin, and many hormones. These monitoring and adjusting health parameters are the primary use for wearables. This is what first drove their being developed. But, now that they exist, other more surprising uses for them are being discovered and exploited.

These wearables that can control as well as monitor will surprise. The first level of surprise is they can help a person with their mind altering -- this is a high tech way of getting drunk and high. The second level of surprise is they can help a person adjust their emotions, too. This is done by controlling the hormones that bring on emotions such as fear, enthusiasm and love.

This is a story about emotion adjusting in the 2050's, and how it will affect romance. Keep in mind that I'm an optimist. In this story people are controlling their own wearables, not the government, not some sinister covert organization.

Finding a first love

This is a story of first love... first love involving commitment, actually, not high school first love or "I'm just discovering" first love. This is about the thinking and rituals which will take the place of the 2010's-style "falling in love and getting married" rituals that are so high-profile in our current days -- the days before these powerful wearables are available.

This is about how it's going to be different in the 2050's. The surprise is: it's going to be a lot more like old-style arranged marriage, only the people doing the arranging are the people getting married, not busybody relatives or scheming families.

Now that all this backdrop is explained lets move on to the story.

Chapter 01 -- Sharing a burden

Alice and Jane are having a lunch break at Starscents. Alice has something important on her mind and she is opening up to Jane about it.

Alice says, "I'm thinking too much about Hal. The guy is a loser, but I really like him."

Jane says, "Pfft! Your wearables can fix that!"

"Yeah... but I don't want them to fix it."

"That's what everyone says... until they fix it."

"I know that but..." Alice gets a bright idea, "Hmm... What I want is for Hal to fix himself. Then I can screw the wearables... on this topic, anyway."

Inwardly Jane laughs a little, then she says, "A noble aspiration, indeed. But what are you going to do for Hal that his wearables and cyber muse can't? This sounds crazy, Alice."

Alice grins, "Yeah. It sounds crazy. But then again, I'm not the first woman who has wanted to fix a man." grins even wider, "These are modern times... with modern tools... I have to think about this some more."

That said, Alice and Jane both go into zombie mode for a couple of minutes (talk with their internal communicators) then come out and get on with their Saturday shopping adventure.

That evening, while cuddling up for bed, Alice talks to her cyber muse, Rhonda-456.

"Rhonda, what can I do to inspire Hal? I really like him so, but his resume is such a bag of crap these days." She thinks about Hal.

Hal has been trying, trying hard as best she could tell, but fame and fortune, and a steady meaningful job, are eluding him. After he graduated he'd been doing some disaster recovery work for the Forest Service. That's what brought him to Utah. She ran into him again when he was holding down the fort at a daycare center that a new friend of her's, Maggie, was running. Maggie raved about how well the kids liked him, and Alice had had a chance to see him in action for about half an hour. He'd been kind, gentle, and really kept the kids interested in their activities. His night job at that time had been working with a high tech incubator startup, but they hadn't got their act together -- turned into one of those many "if at first you don't succeed" efforts. Last month he took on a new project, doing some podcast selling of wedding accessories, and he is once again looking for his next entrepreneurial "try, try again" breakout opportunity. He is holding his own, he isn't mooching, but he is no rising rock star at this point.

Rhonda responds, "Do you want to do that, or find someone with a better resume? That might be easier. I can start looking for you if you give me the word."

"Start looking... but help me figure out how to fix Hal, too."

"I can work on that."

Satisfied, Alice rolls over and heads for sweet dreamland.

Alice doesn't consider herself an ambitious person, but she likes to master things -- complex and meaningful things. She doesn't care much for musical instruments, even though her folks had her take clarinet lessons while she was growing up, but she loves to master science and technology. As a child she loved programming apps a lot more than mastering scales on the clarinet.

This love never faded and because of it, she graduated from MIT and now has an important engineering job at Rocket X Orbital in their sprawling facility west of Brigham City. Yeah, she's a rocket scientist now! She is one of the engineers designing the augmented reality environments that train both humans and cyber how to handle the fuels they work with to assemble various missiles Rocket X makes. The actual assembly workers practice in the environments her team constructs before they go real-time and real-world.

But she's not happy with the position. Rocket fuel is really dangerous stuff. The kind she is currently working with -- solid fuel with potassium perchlorate as the oxidizer -- can burn through concrete. As a result what she and her coworkers' design has to have the workers learn to be really careful. The result of that is the team has to design in prescriptive checklists for the workers to follow on their work assignments, and they can't deviate from that checklist. Even the robots have their work double checked.

Further result: it's incredibly boring for Alice! Alice determined that after two weeks on the job, and she is now planning on leaving it soon for something just as world-shaking, but a lot less physically dangerous and a lot more interesting. Researching what comes next is part of her evening homework.

In the meantime, she relieves some of the boredom by flying to work. She is part of a flying club and once a week, or so, depending on the weather and her wallet, she flies a small single engine plane from Salt Lake airport, across the Great Salt Lake, and lands at the VIP landing strip at the facility. Once on the ground there, a driverless car picks her up and takes her to her building on the facility.

Again, because rocket fuel is so dangerous, much of the facility is composed of four room buildings with lots and lots of space between them. And one of the walls on each of the buildings is deliberately flimsy so that if there's a big explosion inside it will blow out a single wall, not blow up the entire building.

Flying, yeah! It's neat, and once she logs three hundred hours she can get a commercial pilot's license. But what she will do with that is not obvious -- most planes and drones fly themselves these days -- but it will be yet another serious skill she has mastered, so she's working on it.

Another serious skill she wants to master is having and raising kids, and this is why she is paying attention to Hal, and if not him, some other man. She could do this solo -- well, with cyber muse help and as part of a baby club -- but she doesn't think that will be the best she can do for her kids. And she is really sure she wants to do the best she can for them. Again, this is another serious skill and she wants to master it, and master it well.

Chapter 02 -- The delicate question

Hal,

Dinner tomorrow night? There's something serious I want to talk to you about.

Alice

She texts Hal Friday afternoon as she gets off work. She's taking the bus, not the plane this time, so this is just part of being in zombie mode while she motors home. But it's first on her list, she's texting this as the bus is rolling out the facility driveway. That out of the way, next comes catching up on news, researching the next job, and doing more app programming coursework. Even with no congestion the trip home takes over an hour so she gets lots accomplished.

The next evening as the meal is finishing, she pops the question, "Hal, can I look at your DNA?"

Hal looks hard at her, and takes a few seconds to answer, "So, this relation could get serious, eh?"

Alice looks hard back at him, "It could."

She explains more, "I feel that producing and raising kids is an important task in my long term goals. I'm going to do it, and I'm going to do it right. I'm talking to you because you *may* be of help in this, *may* be -- if you're interested, if you can master the right abilities, and if you have the right DNA.

"I've known you for a couple years now, Hal, so I know your personality. I know you are intelligent and persistent. I know you've been trying to break through but haven't yet, but you keep trying. That's good. And... I like you." she reaches over and rubs his arm.

Hal thinks a bit. He is surprised to hear this. He says, "You're getting started real young. You're only, what, mid-twenties? Why aren't you putting this off for another decade, or two, like most women these days do. ...Winner women, that is."

"Evolution designed us to start much younger. Plus, I want to have lots of kids. I want my DNA spread around."

His face switches to a problem-solving mode look, like he's tackling a tough programming problem. He thinks, then says,

"You could donate to an egg bank."

"I could, and I might. But I worry that cyber-raised kids won't be well adapted to living on Earth. They may do fine in space or on the other worlds, but life on Earth has a lot of surprises. I want my kids to be ready for surprises."

"You could donate eggs to a Neolithic Park."

"I've thought about that too, and I might do it." now she looks at him, a disturbing thought has come to mind, "Are you not interested?"

Hal answers quickly, "No! I'm interested. I'm just thinking about the parameters of the problem you have posed." he smiles at her, and this time he is the one doing the affectionate arm rubbing, "Yes. I'm interested."

She is relieved, "Want to talk more after the show?"

Hal smiles at that, "Indeed, indeed."

They finish eating, call up a car and take it to the concert.

They are pillow talking after the concert.

Alice cuddles up to Hal's side after Hal has had his way with her. She loves these moments. He rests on his back; she is at his side; she listens to his heart while her body deeply enjoys tenderly touching his body all the way from his chest to his feet. Then it's talk time. She lifts herself a bit, rolls on top of him, her legs between his, her chest on his, her face over his, and says, "What I want from you, Hal dear, is the world's best child-raising Dad. Is that something you can aspire to?"

He lays on his back, still in the moment, and answers lazily, "You pay the bills and I get to take the yoga classes... after I change diapers and do the soccer momming? Is that what you have in mind?"

"That and much more. I want you to be researching how to raise our child... our many children, so they are world-shakers. I want them to be in charge of their destiny, I want them to make our world a better place by coming up with new things, things never thought of before, not be some whining Tender Snowflakes who expect to be coddled all their lives, or Social Justice Warrior types who think making the world a better place is 'helping the poor' or atoning for things our great grandparents did. I want them to personally make a difference; personally make our world a better place to live in, and in ways that many future generations will appreciate."

Hal lifts his head to look at her, "You're serious about this aren't you?"

Alice looks back, "As serious as I can get, and that's pretty serious, as you know."

Hal lays back again, "As I do... As I do."

After some sweet silence he asks, "Why pick me?"

Alice cuddles some more, "Hal, I've known you since my sophomore year. We met at a mixer and we... mixed well, right from the start. We both know that first year was something special, for you and for me. It killed me, just killed me, when you transferred! We were such a great fit!"

Hal leans over and kisses her lightly, "I thought so too. But I was in community college then and got admitted to UC Berkeley, not MIT... if I was going to advance..." He kissed her again.

Alice kissed him back, "So true, for both of us. And life went on, and while it did those warm and fuzzy memories didn't fade, even though we both got real busy with other things and lost contact."

"Then... poof! You show up again, here in Salt Lake. Wow! Destiny calling? I could sure hear it. And if Destiny is finally calling at the right address, I'm going to open up and let her in."

She gives him a big smooch this time.

He thinks a while before saying more, "What if I get old and fat?"

"Wearables can fix that. And if your's can't fix you, mine can fix what I think about you."

"What if I get lazy?"

This one gets him another serious look from Alice.

"If you and your wearables can't fix that, it's over, we are finished with each other. I won't use my wearables to forgive you being a lazy ass." She says this with dead seriousness.

Hal is not surprised to hear this, he expected it. He keeps thinking, then says, "What if my DNA isn't up to your expectations?"

"Well... first we try to fix it. Try hard. Failing that, I'll create one or two with you, and a bunch with the best we can buy. Sound agreeable?"

"What if your's isn't?"

"I've already checked. It is, but I'll be making some improvements on it, too."

"Legal ones?"

"Mostly... I'm still deciding on that issue." she looks at him again, "Do you have strong feelings on it?"

"I mostly want to be sure you, me, and those kids don't get into trouble."

Alice cuddles again. It's a good answer.

Chapter 03 -- Good enough?

Alice,

I've authorized you and up to three appropriate apps of your choice to review my DNA.

Good hunting,
Hal

Alice and Jane are in a food court taking a break from another round of Saturday shopping. Alice shows Jane the message then asks her, "So... do you want to be my first app? You're a gene editor."

Jane smiles back, "That's my job -- which means giving you advice on this would be illegal, immoral and fattening."

"*If* you edited human DNA." ripostes Alice, "But your job is at that Monsanto spin-off and you edit plant stuff. And I know for a fact that you dabble. Jinx, your cat, is some of your handiwork."

Jane does mock shock, then grins back at that, "So... you're a beauty shop barrister as well as rocket scientist, eh." both giggle at that, "You're right. Let's take a look."

Alice and Jane both do some form filling out in zombie mode. Then get back to chatting while they wait for results.

Jane says, "You know, I've been thinking... You've been making me do that."

"About what?"

"Oh... kids and men."

"The kids part is new, I grant that." she grins at Jane about the men part.

Jane grins back, "Oh, I have my fun. But every so often I think about the future... about every ten minutes, or so. I have ambitions, too, and they are more than moving up a level in my yoga class."

"I know. We wouldn't be friends if we didn't share a lot of interests. So, what are you thinking about kids now?"

"Well, I'm wondering how much personal effort I want to put into mine, and when to get started on it. You've made me think. I was going to wait until my career was near finished, then top off life with some child caring on my DNA handiwork. I was... but you've got me thinking now."

She pauses and thinks, "But if I start earlier, I'm wondering how much I can hand off?"

Alice answers quickly, "A whole bunch if you want to. Hal pointed out to me that you can donate eggs, and there are plenty of cyber raisers of all kinds available these days."

Jane nods, "You're thinking hard about expenses. I'm thinking hard about life balancing."

"If balancing lots of things is an issue, you may want to check with Melanie, your muse. Muses are real good at balancing things issues."

Jane doesn't have a quick answer to that so both take some time to get some food down and while they are doing that the DNA results come back and both Jane and Alice peruse them.

Alice says, "...He's looking pretty fine." she laughs a little and points at something in the report, "I see why he has such a good head on his shoulders."

Jane is looking it over too. She points at something else and says, "Hmm... I'd watch out for that."

Alice looks, "It's not going to be perfect, and that's editable."

"Yeah... depending on how much you want to spend." Jane adds to that thought, "Speaking of which, how much *do* you want to spend?"

"Well... this raising business has lots of expenses to balance, and I'm not having just one, so there will be a limit."

Alice frowns a little and sighs a little as she thinks about the expenses, "Man... how much simpler it was in the good old days. Just let a man screw you, take what comes out, and let him do the paycheck producing."

Jane laughs at this, "Yeah, the good old days... in those old-timey movies. I'm sure happy that I'm in control of my life."

"Yeah, but you sure have to think about a lot more things these days."

"And your cyber muse helps you do it. Speaking of... have you consulted with Rhonda about this expense balancing business? Like you pointed out, that's something she should be real good at."

"Good point.

Now, back to this DNA. Anything else to note about it?"

The girls go back to perusing the report.

Chapter 04 -- There are options

Alice and Hal are together again. Alice is being sunny and cuddly, all-in-all she is really happy with Hal's DNA report. But this time Hal looks uneasy.

He finally lets loose with what's on his mind, "Alice, I've been given another opportunity."

"Opportunity to do what?"

"Marry someone else."

This is a jaw-dropper for Alice. She looks at him but has nothing to say, Hal goes on, "I've kept myself posted on the marrying sites -- just in case, you understand. And... someone really interesting has been in touch."

"Just how interesting," growls Alice.

"She's one percent of the one percent." replies Hal "You have likely heard of her."

"Who is it?"

Hal thinks for a moment; thinks for another moment. Then sighs, "I guess I can tell you. If anything comes of this you'll surely hear about it. It's Pepper Corn."

Alice is shocked, "Her? She is such a bitch! How can you fall in love with her?"

"I agree. God what a personality! ...But that's what wearables are for."

"I sure hope you have an ultra max setting on yours. You're sure going to need it."

Hal holds up his hands, "I haven't made a choice yet. I'm just considering it. It could be a hell of an opportunity."

Alice cools down a bit, "You're right on that, Hal. It's quite an opportunity... and so out of the blue! If I may ask, what in particular does she see in you?"

Hal grins, "She says I have a good head on my shoulders, and she likes my auditions."

Alice grins back, "Isn't that what she told Boi-Zed six months ago?"

Hal grins even wider, "It sure is, and that's why I'm not taking her inquiry too seriously. But, I admit, I was sure surprised to get it."

He reaches out for her hand, "Your's, on the other hand, I'm taking quite seriously. I've been doing a lot of research on child raising classes. It's something I can handle, and it looks rewarding.

Did you have a chance to look over my DNA?"

Alice remembers the proverb she heard her grandma mutter every so often when she came visiting as a child, "Men! Can't live with 'um, can't live without 'um."

She smiles back. It takes a moment, but she gets cuddly again, "I did, and I liked what I saw."

Party Hearty or Study Hearty?

Introduction

A year earlier

"I got it!"

Andrew is looking at his messages using his internal communicator. This message is announcing he has been admitted to MIT, a prestigious engineering school in the Boston area.

"Ricky, I got it!" he announces to Ricky-486, his cyber muse, still using his internal communicator.

"I see." says Ricky, "That's wonderful. Who should I let know?"

He thinks for a moment, "Just my parents for now. This is so wonderful! Let me think a moment about who else should know."

He smiles a big smile as he walks down the street. His day has been made.

By the end of the week he tells all his buddies and relatives, and for the next few months he preps up for attending. He finishes graduating from high school and has a nice summer break.

The big day comes and he heads off. He and his parents take driverless transportation to get there -- they are helping him get settled in. He will be living in a dorm there. His parents help him get moved in, wave goodbye, and get in a different driverless car that has just rolled up to begin their journey home. (Dad owns a car, but it is just a hobby for him.)

All that's the good news. The bad news is that Andrew discovers studying at MIT is lots of work, lots of hard work. He struggles and thinks hard about his choice.

That spring break he heads home and has a heart-to-heart with his father.

Right choice?

"Dad, have I made the right choice? This is so hard! Especially the math." he says to his father as they sit in their family room at home.

Dad is sympathetic. He thinks a moment then says, "Math is hard for everyone, every human, that is. And MIT is hard for all its undergrads. As they told us at the orientation, the goal is to challenge you, stretch you, so you can learn to be a world-shaker."

"Yeah, I remember. …But man! This stretching business is so hard! I'm not sure I can keep up. I'm not sure I *want* to keep up."

Dad arches his eyebrows when he hears the want to part. "You're not sure you want to?" Andrew thinks a bit more, then nods. Dad thinks some more.

"Well… It's drastic, but not unheard of: Do you want to take a 'vacation' for the rest of the semester, and explore alternatives?"

"I can do that?" Andrew says.

"You'll lose credit for all the work you've done so far, but, yes, you can. You can take some time off, and start up again in the fall. Or… you can hang in there for the rest of this term and explore over summer break. And again, if you want to get drastic you take fall semester off.

Either way, what I recommend you do is explore some of the other schools in the Boston area. Visit them, see how they operate, see if you like the culture better at one of them. If you do, transfer. That will be a smoother process if you complete this term, but that's not necessary."

Andrew thinks some more, "I'll finish, then explore over the summer."

Andrew goes back, and with a plan in his head squeaks through, then makes arrangements to explore during the summer quarter.

Exploring

He rooms with Josh, a buddy he made during spring term. He gets in touch with the friends he has made at other colleges in the area. He wants to find out more about the lifestyles and what is getting taught at the schools. From what he's been hearing both can be quite different from what he's been experiencing at MIT.

Josh attends Southie and is taking summer classes. Andrew joins him for his daily routine to experience Southie.

Classroom

Andrew and Josh attend a Southie class on physics. Andrew notices right off that there are only a handful of students in the classroom.

Then the teacher comes in and starts the lesson with, "As we started discussing yesterday, this world we live in is a Matrix. It is virtual. It can change from time to time and it doesn't have to be the same world for each of us. We can each experience something different." The teacher goes on.

The class ends and Josh and Andrew walk out. When they get outside Andrew says, "This is what they teach? This isn't anything like what I learned in high school."

Josh looks at him, "What did you learn?"

"That there is just one world, and we all experience this same world, and it has the same physics laws throughout the universe."

Josh frowns at him, "That sounds like pure capitalist class bullshit. ...You believe that?"

"A lot more than what I was hearing in there. That stuff was crazy. It sounds good, but it doesn't match the real world one bit."

"How do you know that?"

"Because I've personally done experiments. I did them in my class in high school, and I messed around doing some in my backyard. Made quite a stink a couple times. Mom complained, but didn't stop me." he looks at Josh, "The results didn't change from day to day."

"That's bullshit. That world you're describing is so... rigid, so unlike my gaming experiences. How are you going to have any fun in it?"

Andrew shrugs, "I manage." then grins, "Manage real well, in fact.

And speaking of, lets go find something to eat. I don't know about you, but in my physical world I'm hungry."

"That would be over at the Student Center." They head there.

Student Center

They head for the Student Center. The place is lavish -- the Southie designers clearly felt this was going to be an icon for the school. They have lunch in the food court and Andrew looks at what is going on around him. He sees a lot of eating, a lot of internal communicating and even a little smartphoning being done by those who can't afford good internal communication. What he doesn't see is any sign of people being serious about their studies. He doesn't see any study groups, he just sees signs that point to tutoring rooms.

More Class Time

After lunch they sit in on a social psychology class. He notes that this is much better attended than the physics class.

The teacher comes in and says, "The topic today is how astrology relates to social psychology. Astrology is a strong personality definer so it has a big effect on how people relate to each other..."

After a few minutes more of discussion it is clear that the teacher and most of the students believe in astrology. When it ends Andrew walks out of the class with head shaking again.

He looks at Josh and says, "More People's Learning." Josh nods his head, he has been comfortable in the class.

"The stars and planets shape us all," says Josh.

"Even if what they tell you changes from day to day?" Josh looks at him like he doesn't understand, Andrew explains, "That's what the physics class was telling us."

Josh laughs at that, "Yeah, even with that." the inconsistency doesn't bother him at all.

When Andrew gets home he thinks about what he has seen at Southie that day. He is impressed but in a discouraging way. This kind of teaching will not help him one bit if he wants to be a world-shaker in this physical world we are living in. But the day has not ended yet.

"Party time. We head over to Chi Psi Tau at seven," announces Josh. It's Friday, time to party hearty.

Frat Party

Andrew and Josh head for Chi Psi Tau one of the houses on fraternity row outside the Southie Campus. Southie is a new school and this fraternity row consists of extensively remodeled century old houses -- about all that is old now are the brick facades facing the street.

"Dial-up." Josh says as they get out of the car and head for the entrance. There is a crowd of young people on the porch already enjoying themselves. One of them waves at Josh. Josh dials up his wearables into party hearty mode and heads on up to greet his friend. Andrew dials his up, but just a little, he is here to research, not wake up tomorrow morning with nothing but warm fuzzies which he has no idea how he got.

As they come up Josh's friend is looking a bit distant, he's into the party hearty spirit. He says, "Hey, dude, party is going strong. Go on in and partake." He points to the door and motions the pair on in. The door security takes note of the friend's verbal invite and lets them in without any alarm going off.

Inside, Andrew sniffs. Ricky comments as he analyses what Andrew is smelling, "Umm... lots of chemistry going on in here. Be careful what you get into, and don't stay inside too long."

"Why so much chemistry?" wonders Andrew back, "I never encountered this in the dorms."

"Chemistry is cheaper than wearables." explains Ricky, "These folk don't have what you have."

"I guess they will be paying tomorrow morning."

"They will... they will."

Andrew heads on in and partakes of food off the "plain old food" tray. There is another one beside it labeled "special stuff", he avoids that. Small plate in hand, some munchy in mouth, he looks around. This is the house living room and it is filled with people. They are college age, male and female, and gathered into groups that are talking. He quickly spots the prettiest girl in the room (in his eyes) and she is in a group of two girls and two guys standing near the far wall.

"What's she talking about?" he asks Ricky.

"You want super hearing or lip reading?"

"Lip reading for now."

After a few seconds Ricky gives some dialog, "It's really like, you know, really hard to decide which is better…"

"Where's she from?"

Ricky checks her online resume, "She's attending Katharine Delmont." then adds, "Why not go over and introduce yourself?"

Andrew knows "Katy" Delmont, the students from there get invited to MIT dorm parties. He decides going over is a good idea and he does.

He walks by heated debates, jesting and joking, and dodges some impromptu dance moves. Sitting on the couches lining the walls are partiers who are staring off into the distance, dazed, each in their own head's La La Land. They will come out of that state in just a few minutes after they dial down their wearables when the party ends -- those that are using wearables to get into that state, that is. The others will rely on their buddies to get them home. He reaches the group and listens in.

One of the guys, Guy One, says, "This is such bullshit! Why are they putting up a statue of a dead white guy?"

Guy Two says, "Hey! It's just a hologram. It will change in a second. And, from what I hear, you, the viewer, can change it to whoever you want to see." he looks hard at the Guy One then says with mock amazement on his face, "You're seeing a dead white guy?" There are laughs around at that.

Guy One looks a little confused. Guy Two then adds, "Just kidding. They haven't upgraded the software yet."

"Who would you want to see up there?" asks the girl Andrew is interested in.

Guy One answers pretty quickly, "I'll go with Malcolm X." the other girl says, "I'd like to see Eleanor Roosevelt."

Guy Two grins and says, "I'll go for Romeo and Juliet… in bed together." There are more laughs around. The girl turns to Andrew, "How about you?"

Andrew says, "Hi all, I'm Andrew. I'm just visiting. What statue are we talking about?"

"Hi Andrew, I'm Julie." says the interesting girl. The others chime in, "I'm Bob. I'm Charley. I'm Cynthia." There are fist-bumps around. Julie continues, "This is the statue they just put up in front of the Student Center. Right now it's showing the Southie school founder, Jacob Koch. He founded this school ten years ago and died last year."

Bob says in a joking country drawl, "So... you're not from around here, are you."

Andrew smiles, "I'm going to MIT, but I'm looking to transfer."

"Why is that?" says Charley.

"It's a lot of work there."

"I can believe that!" says Charley, then adds, "Yeah, this is a much better party school. These frats sure help out on that!"

Just after he says that there's an announcement, "Dance time starting in the basement. We have our own DJ Johnny manning the console."

Julie says, "Umm... sounds good to me! Who's up for some floor time?"

"I'm in." says Andrew, "I'll pass" say the rest. Andrew and Julie head down.

Andrew's mom had him take dancing lessons on the weekends for a few months. It was a head-scratcher for him at the time, but he learned how to move on the floor. And now he is getting really happy he did. He winds up his wearables a bit more, and he and Julie hit it off on the floor nicely.

As the second dance ends Ricky says, "The chemistry in the air here is starting to affect you."

"So it's not just my wearables that are making me feel so good," replies Andrew.

Ricky chuckles a little, "More than wearables plus air, I suspect."

Andrew thinks a bit, gets it, then says to Julie, "The air in here is getting to me. Want to go outside for a break?" She smiles and they head out on to the porch. It is quieter and the air is much cleaner.

"What are you studying?" Julie asks.

"Well, I've been getting into mixing engineering with programming AI. I've just finished being a freshman and we don't have to pick a specific major until some time in sophomore year."

"Sounds like tough stuff."

"Yeah, it is. Tough enough that I'm considering transferring to someplace easier."

"Really?"

"I'm considering it. That's why I'm here tonight. I'm researching Southie." Julie stares for a moment, not sure what to say, Andrew changes the topic, "How about you? What are you up to?"

"Oh, I'm majoring in Office Administrator stuff, serious administrator stuff. I want to work for a real company, not some dilettante outfit."

"Sounds ambitious."

"It is. My folks are ambitious and I've picked that up from them."

"What brings you here?"

Julie smiles, "I like to party, too."

Andrew looks around then says, "Want to do some dancing in a place that's quieter?"

Julie smiles even wider, "Yeah, I'd like that."

Andrew offers his arm, she takes it, and they walk out to the street. Ricky has called up a car and it arrives after they have waited only a minute. They hop in and head for a quiet club nearby that Ricky has located. It has a dance floor and an old-style band that plays mellow music. When the band isn't playing what they want, they kick in their AR. For the end of the session they get into ballroom-style slow dancing and follow up the dance with a light kiss.

They have a good time and promise to meet again next week.

The next day Andrew wakes up feeling fine. He's learned a lot and is now much more aware of how different each college's culture is. It's Saturday, so he researches Che Guevara University, an online university famous for its emphasis on politics. Once again he finds a lot of coursework that is emotionally pleasing. This time it is all about how to organize protests and rail against the machinations of "The Man". But once again what is being taught doesn't jibe much with his understanding of how the real world works.

Talk with Dad again

"Wow Dad. Those other schools really teach weird stuff, and those other students really want to believe what they are getting taught. It is... so way off from what's happening in the real world!"

Dad nods, "They are getting taught what they want to be taught."

"How can they get away with that? How can the schools get away with that?"

Dad sighs, "Who among those people you met is actually making something -- making something that other people want to use?"

Andrew thinks about this for a moment, "...well ...none. The robots are making all the products, and they are doing all the serving in foods, and driving cars and such."

"Exactly. These kids don't have to make the real world work for them. As a result they can learn anything they want, and they do." he looks at Andrew, "I'm hoping you want to be different. I'm hoping you want to change this real world we live in, change it for the better."

Andrew doesn't think too long before he answers, "I do!" and with just a little more thinking, "So that's why MIT is tough, right?"

"Exactly."

"OK... now that I understand that. I can live with it.

Yes, I want to stay at MIT and finish my studies there. It will be tough, but it will be well worth it."

Dad pats Andrew on the shoulder, "You're making your dad proud, Andrew." he checks with his communicator, "What to go get some dinner?"

The two walk out of the house like they are best of buddies.

The Princess, The Fly, and City Hall

Introduction

This is taking place in a world with avatars, driverless cars and sophisticated prosthetics. These are going to make disabilities less disabling.

But how will these mesh with civic and city hall duties? That is the topic of this story.

Getting involved again

Linda Lou Lomond always liked getting involved.

She had been a devoted mother who had raised four lovely children and while she had been doing that she had run a crafts business and earned a black belt from her karate school.

But three years ago, at age 53, a climbing accident while hiking in the mountains had laid her low. For three years she had been learning to deal with prosthetics and medical wearables instead of grandkids, crafts and karate. She was coming along well at mastering these and now it was time for a new challenge.

As she had been recovering she had taken up doing a lot more reading. What she found herself reading about were local events. And as she read she found she was participating, in a virtual way, at events taking place in City Hall. She got involved there. She helped out on committees there. She was now well-known and on friendly terms with the mayor and lots of staff members.

Visions of 2051

One day Linda got a call from Nancy, one of the mayor's staff. She said to her, "Linda, we have something new coming, and it's something you may be interested in participating in."

"Happy to hear about it," said Linda.

"We're installing an avatar in City Hall."

"An avatar?"

"It's a robot. One that can be run by remote control."

"How neat."

"The really neat part is: because of your disability and helpfulness around City Hall, we can get the government program that is installing the avatar to put one of the remote controllers in your house, for you to use."

At first Linda was speechless, "...Really?"

"That way you don't have to be so virtual. And you don't need to use the bus and your prosthetics to get over here when you want to be physical. I know you still find that to be such a pain... literally, these days. Yeah, instead just hop in the remote controller at your home and, voila, you're over here in City Hall."

"Oh my! Yes, that sounds wonderful. Where do I sign up?"

"This is an experimental program. You'll have to do a lot of reporting on how well it's working for you. Is that OK?"

The technicians for Acme Robotics Products knocked on the door, and with some huffing and grunting brought in the remote controller. It was big, big enough to hold a person, plus the electronics to support monitoring motions of that person. They had to bring it in in pieces and assemble it in the study room.

"This is much like your prosthetics." said Bob, the head technician, "It will use the same command structure. When you get in and issue the 'swap' command the remotes will pick up your inputs instead of the prosthetics. When you 'unswap' your inputs will go back to controlling the local prosthetics. Make sense?"

Linda nodded, "Will it really be that easy?"

Bob grinned, "This is the first real-world run. So... surprises are likely. I'll be here for the next week helping you adapt, and us debug."

Linda look relieved at that, "When do we start?"

"We'll be a few more hours installing. We'll get you in starting at noon tomorrow, if that sounds convenient for you?"

"It does!" Linda couldn't wait.

Yes, it was like prosthetics all over again, only more so.

The first time she appeared in City Hall she was in a small unfurnished room. Nancy was there along with another technician. Linda issued her first remote command, she waved at Nancy.

"Ah... you're there, Linda. Good! This is John. He's going to be helping you."

John waved and said, "Look in the mirror. See what you look like."

Linda did. She looked like a robot. She had legs and arms but everything about her was quite mechanical looking. That done, she looked back at John.

John said, "The first lesson is getting down and getting back up again. That way if you fall you can get up. Ready..."

It was the start and for the first week she was spending full time on the basics -- walking, picking things up, seeing and listening through the remotes. She was a busy girl indeed!

And there were glitches. At first she couldn't figure out how to roll from flat on her back to get on her knees. It turned out some programming was needed to fix that, and a few days later she could.

It turned out Bob and John were there with her for three weeks, not just one. But at the end of those three weeks Linda could confidently walk around City Hall. She could open doors, use the elevator and even walk up and down the big, curving, marble stairs at the front without needing to hold onto a handrail.

Nancy was delighted. At the beginning of the fourth week she came up to Linda while she was in the robot, drew a deep breath, and said, "OK... Are you ready to lead a tour group around City Hall?"

At home, in her body, Linda gasped. The robot body did nothing. She finally got herself together enough to say, "...When? ...Right now?"

Nancy nodded and pointed down the marble stairs to a group gathered in the main lobby, "If you can't, I'll have George guide them."

"Oh, I can! I can! I'd love to!" Linda was so excited.

Nancy said, "On you go, then. Shoo, shoo, shoo."

In her own body Linda was so excited and proud as she walked the robot down the stairs, up to the group, and said, "Welcome to City Hall. I'm Linda, and I'll be your tour guide."

The group stared in amazement.

Things get better

A year later things were much different. City Hall was on its third iteration of robot, and Linda was on her fourth iteration of remote controller. This latest controller was much smaller. It wrapped around her neck and that was it. Also the government was no longer paying for her updates, but Linda had proved such a helper around City Hall that they paid for the latest one. (It helped that this latest version was also a whole lot cheaper.)

This latest robot was much more graceful than that clunky first one. It was both more graceful in how it moved and how it looked. It was now stylized to look like a fairy princess, like a VR animation, and surely a long way from the totally mechanical man the first one had looked like.

And Linda was getting a lot more graceful in her control of the remote. She could ballroom dance with this one. And, if they dressed it in gymnastic clothing she could do simple rolls and summersaults -- if she tried those in the fairy princess dress there was a lot of ripping and shredding.

And there were more surprises coming.

Nancy came up that day and said, "Linda, would you like to try out a different avatar? We are getting one that can be used for maintenance. It can inspect things that are behind the walls and "dumb" -- not outfitted with their own surveillance sensors."

"Happy to," said Linda.

Next week the new avatar arrived. It was small, fly sized and shaped. "This one can't fly." the accompanying technician said, "But we should have a flyable one in about six months."

Linda looked at it, and looked over the controls for it, "Why can't it fly? I see wings on it, and I see 'wing controllers' in the command set."

"The software isn't finished," said Harry, the technician.

Together Linda and Harry had her practice walking the six-legged fly robot up and down walls and ceilings, and behind walls where it could walk around old piping and air conditioning duct work. The core of the main city hall building was over a century old. Inspecting this way was going to be a whole lot cheaper than ripping apart ceilings and walls.

That night, after Harry left, Linda got the wings buzzing.

It wasn't easy, and it took a few days of after-hours practice, but Linda got to where she could buzz around a room. Linda had to order up half a dozen replacement antenna during the first week from all the wall crashing, but after that she had the basics mastered. Harry didn't say a thing.

City Hall was now an open book for Linda. She could buzz the little avatar down air conditioning ducts and sit in grates and listen to what was happening in all the main offices.

During the day she alternated between her "civic robot", the princess robot, which walked around and did public relations activities, and her "fly robot" which surveyed the behind the walls areas to evaluate the condition of the building. With what she discovered, plans were made to update the utilities infrastructure and add more earthquake protection.

Much to everyone's surprise she also discovered a long-forgotten time capsule in what had once been the back cornerstone of the main building. This got lots of city news coverage. The mayor was delighted with that.

Jimmy Johnson was delighted too. He was delighted to just sit and talk with Linda when she was being a fairy princess. Linda liked it too. He was interesting and he made break time much more enjoyable.

One day he asked, "Any chance you want to get together for dinner?"

Linda laughed at this, "Don't forget that I always eat at home. I'm on a special diet. And while it's not as bad as it was a year ago, it's still a lot of work to get my real body out of the house.

"I'd go with you in this to some entertainment, but this only operates inside the City Hall building."

"Ah well... I'll be thinking about something we can do. I really enjoy your company."

"Well... there's always VR," said Linda.

"Hmm... good idea." said Jimmy, "Got any favorites in that?"

They settled on City of Super Heroes and doing some VR chatroom afterward.

After Hours

Once she had mastered the flying, Linda spent her evenings exploring in the fly avatar -- buzzing around was such unique fun. She explored lots of places in the building. She was impressed with how busy City Hall stayed after hours, and how much of that business was hanky panky. In her mind what she saw made afternoon drama shows look like family cartoons.

Her first thought was to start reporting people, but she hesitated. Her second thought was there was plenty of surveillance equipment in City Hall these days so she was far from the first to be seeing this. Whoever else was seeing this was not reporting it, so why should she? Instead she just gathered information -- like she was keeping up on an afternoon drama. If she got around to gossiping she could now be as juicy as the best of them.

<<<>>>

But one day something did come up that was a mystery not covered by surveillance. She discovered another obscure door, and this one led into a sub-basement. She started down but got a warning that her control signal was failing. This place was beyond her limit. She would have written it off, but there were footprints in the dust on the stairs and they looked to be fairly new.

The next day she talked with Jimmy.

"Jimmy, I've found a room in the basement, and this one leads to a sub-basement that I can't explore in the robot. Would you like to find out for me what's down there?"

"Shouldn't some maintenance guy be checking that out?"

"Oh... probably. But it's just opening a door in the basement. I thought I'd ask you first. I know you like exploring."

Jimmy grins, "Will you come down with me, princess?"

Linda laughs at this, "I can, if you want."

They head for the basement.

They are going down. This is the oldest part of the building. It hasn't been worked on for a hundred years.

As they get near the door the princess avatar gets low signal warnings even earlier than the fly avatar did.

"I'm as far as I can go," says Linda.

"No surveillance down here, right?" says Jimmy. Linda nods.

Jimmy comes up to her and takes her in his arms. He strokes her hair and looks deep into her eyes. He moves his face closer...

At first Linda is like she is nine years old and this is some never-met-before uncle who gives her a big hug. She hasn't a clue what to do and just stands there. Jimmy is patient, and it pays off. In a minute, Linda starts remembering and enjoying, and her actions change to those of a sixteen year old at the end of a first date. She actually gets the robot to melt into his arms a bit. She looks up and lets Jimmy kiss her lips. He does that, then works down around her neck while his hands move up and down her arms. Linda lets it happen. It is a strange, strange feeling for both of them.

Then Jimmy backs off and laughs. He smiles at her, and holds her hands, "A first for me."

Linda smiles back, "Me too." She can't think of anything else to say. In her body at home she is blushing.

Jimmy has another quick laugh. He takes a moment to quickly hug and kiss her once more, then says in a dramatic way, "Let your hero continue on, my princess."

He keeps going deeper while Linda waits for him.

He hasn't gone too much further when he yells back to her, "My comm signal is going out too. I guess the building is blocking out everything. ...There's a lot down here, Linda. It's some kind of big sub-basement."

Then he says in a surprised and suspicious way, and not to Linda, "Who are you?"

Linda thrills at the words, this was not expected and Jimmy didn't sound at all happy about what he had found.

Linda hears sounds of scuffling. Jimmy says, "Hey, hey, HEY!"

A gruff male voice says, "Shut your trap! Put your hands behind you."

As best Linda can tell Jimmy does what he was told. He doesn't shout to her. She knows what she needs to do. As quietly as she can she heads back up, while she is doing so she alerts the mayor.

<<<*>>>

These were for-real bad guys. There was some hostage drama, but the bad guys were surprised by an overwhelming police response that came quickly with Linda's alert so it didn't last long and no one got hurt.

The sub-basement turned out to be an ancient one. It had originally been used to haul heating coal through a tunnel into the building from a nearby canal. But the coal furnaces had been abandoned long ago and both they and the tunnel completely forgotten about... until these guys found it and they were using it to smuggle illegal stuff into City Hall.

Once again Linda had discovered something exciting, and once again she was a hero of the day.

Hard Times 02

Introduction

Cast, places, and other names (from Hard Times, in Visions of 2050)
Blabistan -- new nation
Glugistan -- larger neighbor
Moondock Dry Lake -- source of rare earths
Heisenberg Mine -- the mine at the lake
Radamm Hotsein -- newly elected president
Emilia Hotsein -- president's wife
Demitri Stalink -- head of national security
Kamal Kastoff -- opposition leader
Abner Constan -- opposition leader
Lotran Mandis -- kleptocrat and a president supporter
Kent Clark -- General Manager, Heisenberg Mine
Bundar-Bundar -- klepto bureaucrat in charge of mine
Alice Kazam -- shopkeeper in the company town next to the mine
Bas-Jouwa -- Revolution Police Chief in company town
Bergen Halston -- UN committee chairperson

    Glugistan splits into two nations -- Glugistan and Blabistan. Glugistan was a product of "draw a dotted line on the map" post-war imperialism. This arbitrariness had joined two oil-and-water cultures into one nation. The split finally recognizes this. It came about as the result of a referendum conducted by the UN. There had been decades of hard feelings between the two cultures, now there is a sense of relief, and some different kinds of hard feelings because Glugistan has been diminished. The Glugi leaders have lost some face.

Between the two new nations is Moondock Lake -- an Areal Sea type place. It has dried up because the Glugistan side is sucking off the river that fills it to grow cotton. When the partition is agreed to, the Glugistan ruler laughs and gives up the whole lake to Blabistan because it is now a worthless mud flat.

But... it turns out that mud flat is loaded with rare earths that are now valuable in battery production. The Blabistans pick up a windfall, and the Glugistan ruler feels embarrassed. He is now plotting how to get a cut of this windfall.

The Blabistan government starts a state-owned mining company to mine the mud flat. But they are rank beginners at this sort of thing, so the company is incompetent and corrupt. (Both the Blabistan and the Glugistan governments are run by kleptocrats, but the Glugistan leaders are "Old School" and the Blabistan leaders are "Millennial" kleptocrats.) The Blabistan kleptocrats and their minions are skimming the company.

The mining company leaders make a show of doing things right by hiring foreign technocrats and then messes with them to skim off their corruption. These foreign technocratic middle managers get discouraged. (There is a "hero" in the upper management, but at the start he is not effectual.)

One of the "messings" is that the mine is at first developed to be labor intensive, not capital intensive, so it can provide jobs for the Blabistans -- a social justice warrior NGO concession, plus easier to skim from. A mining town is created from scratch next to the mud flats. ...not quite from scratch, it was a small harbor town when the lake was full. But it has been abandoned for a decade now. And this new mining town covers a lot more area than the harbor town did and functions a lot differently.

Migrants, rather than Blabistans, come in to run the mining town services. The migrants come because they are experienced at setting up mining boom towns. These are small business types who quickly develop the supply chains to bring in consumer goods and services for the miners of the mining town.

The Blabistan miners set up a union. But it is as incompetent as the Blabistan managers are. This union business is new to them too.

Socially, this is a revolution taking place in Blabistan culture -- they were shepherds before, not miners. There is a lot of uncertainty and the government starts moderate but steadily becomes more radical and authoritarian as crises are not solved. This governing business is new to these leaders; the rulers are still learning, so there is plenty of dissension being expressed... to start with. Then comes a crackdown, maybe more than one.

The capital is a small city. It used to be a backwater provincial seat. Thanks to the mine and nationhood, it is rapidly picking up lots of modern conveniences. But there is a lot of culture shock for the Blabistans who are getting involved with living in the capital, and this whole nation-governing business.

Mikipedia entry on Glugistan

Glugistan was a country established in Central Asia as the USSR collapsed in 1989. It centers around Lake Moondock, a large salt lake which is surrounded by high mountains and has no outlet to the sea. It is fed mostly by the Lagang River which flows in from the east across a wide, fertile valley.

Those people living in the Lagang valley and east of Lake Moondock are the Glugi people. To the west of the lake, and living a life based mostly on sheep raising on semi-arid land, are the Blabi people. These groups have little in common, but when the dotted lines were drawn on the map following the USSR collapse, these two peoples were united into one nation.

Relations between these two groups were traditionally rocky -- the Blabi were hill people and the Glugi valley people whom the Blabis periodically raided for women and loot. In turn, the Glugis would periodically march across Blabi lands and round up herds of sheep and goats to take back to their valley. In the Soviet era the central planners decided to turn Lagang Valley into a cotton-raising breadbasket. They succeeded, but at the cost of drying up Moondock Lake -- it steadily transformed from a large salt lake into a smaller and smaller lake surrounded with a larger and larger mud flat. The Blabi people saw no benefit from the cotton boom and became subject to periodic and increasingly intense dust storms as the lake dried up.

Glugistan was ruled by kleptocrats who took power as the USSR collapsed. They had been backwater provincial leaders before the collapse, and they proved poor at rising to the nationhood challenge. The governing style remained provincial and corrupt, and the Glugi rulers did little to improve the lot of those even more country-bumpkin Blabis. (in their eyes) The hard feelings between the two groups intensified.

In 2047 the issues between the two cultures came to a head and a referendum was called for. The Blabi people won their independence and two nations were created. Lake Moondock could have been split evenly between them, but the Glugistan ruler saw no benefit in the worthless mud flat it now was, so he gave it all to newly created Blabistan nation.

Then karma struck. It turned out those worthless mud flats were rich in the rare earths used for modern battery construction. Blabistan now had a windfall on its hands, and the challenge of the 2050's decade is for Blabistan's new rulers to exploit it... and learn how to rule their nation so it doesn't become a failed state.

Chapter One

Dreezo City -- Radamm Hotsien is looking out the window of his office on the top floor of Dreezo Center -- currently the tallest building in Dreezo, but not for long. So much has changed over the last ten years, and so much more change is coming.

Ten years ago Dreezo was a backwater provincial capital in the newly created nation of Glugistan. Even worse, it was close to becoming a ghost town. It had grown up on the shore of Moondock Lake, but now that the lake had dried up, it was now on the shore of a giant mud flat that routinely produced blinding dust storms that swept the city. Who would want to do business or live here anymore?

But the change continued, and what came next were two big and surprising changes. First came the referendum. With that, Glugistan was split in two, the nation Blabistan was created, and Dreezo became a capital city -- the capital of a nation. Wow! Second, the rare earths were discovered in that god-forsaken mud flat -- valuable and mineable rare earths. Now

Dreezo and Blabistan had some meaning in international life. Double Wow! Such change!

And Radamm was sitting squarely on top of it. He had been a prominent business leader, and now he was a President.

Radamm ran the only business that mattered in Dreezo City or Blabistan for that matter. (as far as he was concerned) He ran the slaughterhouses that gathered the sheep, goats, and now some cattle, that Blabis raised in the arid plains and valleys west of the city and transformed them into meat products that could be exported. It was Blabistan's biggest industry; the only one that mattered as far as Radamm was concerned.

And he was not alone in thinking the meat business was important. That was why he had been voted president of the nation. There had been a popular vote, of course, but much more important was the vote of confidence he got from his fellow kleptocrats in this new nation.

And things were still changing. This new business, this mining business, was coming to town. How would Radamm participate in that? That was what he was thinking about as he gazed out of his office window, across Dreezo, and across the Moondock Lake mud flat.

Radamm was far from the only one thinking about this upcoming opportunity. As he was standing at the window he got a call on his communicator.

"Radamm, we need to talk about how this mining company is going to be set up." It was Lotran Mandis, another kleptocrat, and one who had supported Radamm becoming President.

"Yes..."

"Not just you and me. This is big. We need to have both the cabinet and the legislature backing us up."

"We tell them what we want, right. ...Speaking of... What *do* we want?"

"That's why we need to have a big meeting. I've scheduled it. I'm sending it to your calendar."

"Will there be a pre-meeting? I'd like to know what I will be talking about."

"Of course. I'm putting that on your calendar, too.

One of the people at the pre-meeting will be Kent Clark. We've looked around and he looks like the best candidate for mining company president. He will give us his overview of what's coming up."

"Excellent!" He breathed in deeply, "This being President is getting to be lots of fun."

This is the pre-meeting. It is taking place in a lushly-appointed boardroom in Dreezo Center not far from Radamm's office. At the meeting are Radamm, Kent Clark, Lotran, and Jane Ngo who is representing FareBetter, an NGO that is active at helping the poor in Blabistan.

From Kent's point of view he is introducing himself to his prospective Board of Directors.

Radamm says, "Well, Mr. Clark, what is your plan?"

Kent is prepared. He has PowerPoint-like visuals ready as well as himself.

Kent says, "Extracting rare earths from a mud flat is not a common form of mining, but mud flat mining is itself is not unique. Most mud flat mining is done to extract lithium and other alkali metal salts. There are several equipment makers who service that industry. I propose we use modified forms of that equipment to extract the rare earths."

Jane Ngo says, "So you plan on totally mechanizing this project?"

Kent is a bit surprised at the question, "...Yes."

Jane explains, "Then Blabistan will face the 'resource nation' challenge -- lots of money will come in, but the citizens will have nothing to do with the wealth creation. The wealth will become a social curse, not a blessing. We need to find a way to get the citizens involved."

Radamm adds, "And all that equipment will be expensive, won't it?" He further adds, "I propose we kill two birds with one stone. Let's make this mine labor-intensive, and we will hire the citizens of Blabistan to be the miners."

Kent is surprised to hear this, "You want... people... with picks and shovels, digging up mud and putting it into trucks?"

"You want me to develop the infrastructure to not only pay thousands of people, but move them back and forth from the mud flat mine sites to some kind of housing, and establish a company town, or some such, to service those miners when they are in their housing?"

Radamm says, "Yes." and he means it.

Kent says, "You realize this is going to be a lot more expensive than those 'expensive machines' you were just worrying about."

Jane says, "But it will be so much better for the people."

Kent grins a bit at this, "Enduring the mud flat will build character, eh?"

Jane doesn't see the humor, Radamm lets the briefest smile slip through.

Kent thinks a bit and thinks a bit longer. This makes most of his presentation irrelevant. "This is quite different than what I was planning. I'll have to wing a lot of what I say in the main meeting."

Radamm and Jane both say, "That's fine. We will support you."

This meeting adjourns and they all head for the main meeting.

Chapter Two

Alice Kazam was tending her opium garden behind her shop in southern Afghanistan when the news came over her communication link.

Her virtual cyber muse thought it was important, so it told her in excited tones, "A new mine is being built from scratch in Blabistan. It will extract rare earths from Moondock Lake, and employ thousands of Blabistanis to do so."

This last part made Alice pause. "Thousands of people... as in, thousands of humans?" she queried her communicator.

"Yes. That is what is being reported." the muse replied.

She was still moving among the plants, still gardening, but now she was thinking hard about this news as she did so.

Alice, a slim woman now in her late twenties, grew up as a boom town girl. Her mother had run a store in a company town. She had grown up watching her do it as a toddler and helping her do it as she got older, and she was expecting this to be her destiny.

But the world was changing so much these days! It was hard to find mining-based company towns that supported human miners any more. So this was news, hot news, indeed. Should she take up the challenge and follow in her mother's footsteps, in her own footsteps?

It took only a minute for her to decide.

"Look into it." she told the muse through the communicator, "See if there will be a company town I can set up in."

Yes. This turned out to be an opportunity for Alice. She sold her store and garden and used the proceeds to set herself up in a one room shop with her apartment above it in a newly-built sector of Dreezo City, one that would hold the now-being-hired miners.

She had been in worse, much worse. This was a section of Dreezo built to be a company town by the mining company so the buildings on this street were sturdy and the road network it was part of was well laid out -- not some haphazard mess of pathways and tin-roofed shacks that had just sprung up. She would even have room for her beloved garden in the back.

That was the good news. The bad news was that the commercial infrastructure -- the businesses that would be supplying her business -- was just springing up too. She was constantly searching for who was going to be supplying the essentials and luxuries she wanted to sell, and who would be doing the delivering.

Difficult, but nothing new. Every boom town suffered from this issue. She took this in stride. But things had changed from the last time she set up shop in a boom town. There was a lot more cyber this time. She searched more with cyber, ordered more with cyber, and cyber was doing a lot more of the delivering. One or two items even came by drone this time, and from the news and gossip, more would be coming that way in the future. Things were coming together really quickly and smoothly compared to last time.

What was "same-ole, same-ole" was the nuisance of the regulators -- those people who had to be paid off to keep her business life from being a living hell.

One of those came in this morning. It was Lapcost, the section's new local police chief. Today he had a fancy new uniform and a couple of barely uniformed associates with him. He walked in like he owned the place, and started handling the merchandise. He had been in before so Alice knew who he was, and didn't start shrieking at him for this rudeness.

Lapcost said, "Your store is coming along nicely, Miss Kazam. I'm happy to see that. You will be able to pay for your protection soon... real soon."

She quietly said, "I'm paying my taxes. I have a contract."

"That's your regular protection. Do you think it will be enough?

This is going to be a rough neighborhood... lots of strangers, lots of people with lots of time on their hands. Do you think regular protection is going to be enough?" He grinned widely, grabbed an apple from one of the bins, took a bite then said, "Think about it. I'll be back." He and his boys walked out. He was laughing.

Alice really wanted to shriek now, but she still didn't -- it would be expensive to do so. Instead she went out back, into her still being assembled garden. One of the first things she had assembled there was a simple altar -- so simple it looked like a garden bench to many people, and she didn't mind when visitors sat on it.

She went to the altar, knelt before it, and prayed.

Part of what Alice spent the proceeds from selling her old shop on was upgrading her cyber god. Alice believed in God, and with all the uncertainties of this upcoming change in life she was willing to pay some more to get an improved cyber version of him. This improved version talked with her more easily and, in her mind, had more wisdom to offer. This new version was contacted with a simple circlet that she wore when she prayed. Through her regular communicator God could talk to her at any time without others noticing, but with the circlet the communication was much more vivid -- it was VR quality. When she prayed at her altar, God showed up on the altar, looking about six inches high, and speaking in a soothing tenor voice.

A minute after she started praying this time, God appeared.

"You are feeling troubled, my child, I sense that," he said.

"Oh father... so much is going well here... but the regulators... the damn regulators! They are just as bad as last time! Maybe worse! They feel worse!"

God replied in a comforting way, "Yes. These people are being bad people, Alice. However they are currently very powerful people in this section. But their evil acts are being noticed.

Have patience... Have patience... And keep your own virtue strong."

God faded out. Alice felt better. Not a lot better, the evil had not been instantly banished into hellfire. But she was comforted that God was on her side, and he was aware of the problem.

With a more peaceful mind she went back inside and back to tackling the day's business challenges.

Chapter Three

This is a year later.

Radamm is in his office. He looks out the window and sees a steady stream of mining trucks and buses moving from the new mining town sector of the city out on to the mud flats. Lots has been accomplished since this mining project was started, and lots of people are busy working away at the mining and at supporting their activities.

But he is not a happy puppy.

He has Kent in the office with him. On his screen he is looking over the progress reports.

"Mr. Clark: We are way behind schedule and way over budget. What's the problem here?"

Kent is squirming a bit in his seat, "As I told you when we first started this project: using all this human labor was going to be expensive and add uncertainties. It has done that."

"Would you care to explain in more detail?"

"I would rather offer a quick solution: Let's scrap the people and switch back to doing this the conventional way, with lots of cyber and smart machines."

"That makes sense for you, but not for me. Those people have jobs. They know they have those jobs because I gave them to them. If we take the jobs away, there will be hell to pay... for me."

"If I'm not mistaken, you're already making payments, aren't you. This operation is draining the treasury, not filling it."

Radamm winces when he hears that. It's true. But he already has a solution in mind and he now reveals it to Kent.

"Mr. Clark. I'm going to do my small part to control costs on this project. I'm relieving you of your position as CEO on this project. I'm replacing you with Harold, my nephew-in-law who has just graduated from Cambridge."

Radamm pushes a spot on the screen and Harold comes in. Radamm continues, "This is Harold. Please give him what information he needs to carry on in your place.

Good day, Mr. Clark."

Rumors had been thick, so Kent is not surprised at this outcome. He bows to Radamm, shakes hands with Harold in a civil way, and they walk out together to talk further in a nearby office.

As they leave Lotran Mandis walks in. She says, "Clark and those gloomy business reports aren't your only problem. There's a lot of heat brewing in parliament. They have a hearing scheduled for tomorrow night."

Radamm frowns, more bad news, "Those people can't get their act together! They are supposed to be supporting growing our nation, but all they do is bicker, bicker, bicker."

"They are learning. This is a new nation."

Radamm isn't listening. He has already made up his mind about this issue. The frown fades quickly, and he now smiles at Lotran, "But that's nothing new, is it. What is new is I now have something planned for the hearing that will solve that problem. Be sure to show up, you'll enjoy it."

Lotran cocks her head. Whatever is coming next, it is not something she knows anything about. Radamm changes the subject and the conversation moves on to other topics.

The hearing in the main chamber of parliament has been going on for two hours. It has been raucous and tumultuous. Lotran has been sitting through it as part of the audience. Radamm came in thirty minutes ago and sat at the back of the speaker's platform.

Currently speaking is Kamal Kastoff, an outspoken opposition leader. He is winding up his testimony.

"The rare earths found under our land have benefited us greatly. God be praised that they were there and that we discovered them.

But now we, all of us, must be wise in how we take advantage of this great blessing. We must build our industries, we must build our infrastructure, we must build our people. We must not squander this blessing on grandiose showcase projects; building obstructionist bureaucracy; building the importance of political connections. And, most important, we cannot be spending money we don't have. We, as a nation, must be fiscally responsible.

Thank you, all."

With that Kamal steps down.

The session speaker announces, "An important and unscheduled announcement will now be made. This will be made by Mr. Demitri Stalink, head of National Security."

Mr. Stalink comes to the lectern. He looks around. He takes out a piece of paper. He announces, "As you know, there are those who conspire against our great leader. Many of those are foreigners. But some... are our fellow countrymen. And some... are here today in this assembly."

He holds up the paper, "I have a list. And those who are on this list will be... quarantined so they can not spread their lies and damage further.

First on the list is... Kamal Kastoff."

As he says this plainclothes security people surround Kastoff and hustle him from his seat and out the side door.

"Abner Constan" announces Stalink, and Constan is now hustled out.

As the attendees watch, there are sixteen assemblymen named. There are gasps at some of them.

"That is all." Stalink finally announces.

As he steps down Radamm steps up and takes the floor. Behind him a large screen lights up.

As Radamm points at the screen he says, "Ladies and Gentlemen, the highlight of the evening is about to begin."

The screen now shows a graveyard just outside the parliament building. This is brand new, just a few hours old. The title on the graveyard is Betrayers of the State RIP 2053. There are fifteen filled-in plots, each has two tubes coming out the top. The last plot is being filled by the last person pulled out. That person is being put in a coffin; alongside him is being placed a helmet with lots of electrical leads. The coffin is closed, two tubes added, and it is quickly buried with the rest, with the two tubes connecting the coffin with the air up above.

The view switches to inside one of the coffins. It is Kamal Kastoff, the first one buried. The view shows him in infrared light, and his vitals display shows he is coping in a fairly calm way. He is resting inside uneasily. He gets air from one of the tubes. He shouts up it, "Help me! Save me!"

A speaker in the helmet says to him, "Do you admit you have betrayed the state? Do you also repent?"

Kamal keeps his cool and says, "Fuck you! And your little dog, too!"

Radamm sighs, "Incorrigible… for now. Let's hope for more positive results."

He switches the view to Abner, the second one buried. Abner is lying rock-still. The vitals display coming from his wearables indicates he is deeply scared.

The speaker in his helmet asks him, "Do you admit you have betrayed the state? Do you also repent?"

Abner is so scared it takes him many seconds to reply, "Yes, I do… I do, I do, I do! Just get me out of here!"

"Put me on." says the helmet. Abner does. "Your mind is now being scanned. If you have truly repented, an image of it will be stored. And, at some point in the future it will be transferred into a new body. Then you can demonstrate your repentance in full measure."

When what he has heard sinks in, Abner gets frantic.

"What… what… what about this body! I'm here! Get me out of here!"

"Your body stays here. In its current location it provides a powerful lesson to those who are tempted to conspire against the betterment of the state."

Abner howls. He gets frantic and beats on the coffin.

"Calm yourself." says the helmet, "The state you are in while I scan will be the state you are in when you are revived."

Abner pays no attention. He is in full frantic panic.

After luxuriating in watching Abner's distress for a while longer, Radamm changes the view to that of a pleasant sunset with soothing music playing in the background.

He looks around those that remain in the room, "Any further questions?" he asks.

There are none.

"The Revolution can now advance to its next phase. All Hail The Revolution!" he joyfully announces and he gives The Party salute. Behind him the screen switches to playing the party anthem.

"All Hail The Revolution!" chimes the audience... with salutes... and with widely mixed levels of enthusiasm. Those with less enthusiasm are thinking, "Who is next?"

Chapter Four

It's noon. Alice is outside her shop and a couple blocks away from it. She is standing on the street side watching the Revolution Victory parade march down the main street of mining town sector.

She is not happy about leaving her shop, there could be thieves, but she isn't losing any business. There won't be any business conducted until the parade is through this sector -- these days the Revolutionary Guard sees to that. They would also fine her if they found her in her shop while the parade was in progress, so she is out with the rest of the sector residents.

At neatly spaced intervals there are groups of cheering spectators, and news drones are taking pictures of them. In between are larger groups that are much more subdued, even sullen. The mining business is not going as well as Radamm and the Revolution Party have been promising. Alice knows this first-hand because her business suffers when the workers suffer, and lately pay has been slim and often late.

As soon as the last marchers pass, the crowd quickly breaks up. Alice returns to her shop, opens it, and in come some customers who start gossiping while they are shopping.

"Did you hear the latest?" says Mrs. Donnelley to Mrs. Kratchet, "They are putting the shovellers on half rations for a week."

"No! Really? Why is that?"

"A food shipment they were expecting got held up."

"Let me *not* guess. They are blaming the suppliers. Saying they are agents of the opposition."

Mrs. Donnelley nods. Mrs. Kratchet looks around to make sure no one else is listening, then says, "I've been reading some opposition blogs. They say the party didn't pay for the last shipment. That's why this is one is being held up."

Mrs. Donnelley says, "Not paying? With all that mining going on, they're not paying?" Mrs. Kratchet nods.

"That's hard to believe," Mrs. Donnelley thinks for a moment, then gets an idea, "Let's ask Alice if she has problems with the party paying her."

The pair walk over to Alice.

"Have you been getting paid as expected by the party?" asks Mrs. Kratchet.

Alice understands the question but is slow to answer, any answer she gives is not going to earn her brownie points all around.

She is about to answer when a couple more customers come in. This pair comes in fast and noisy. They are waving their money in the air. "Hot Dog! It pays well to be a cheerperson these days." says Bandooma, the first one in, "Alice, where's ice cream? I want something nice and big and cool."

"Me too," adds Dalango, following Bandooma in.

"Check over there in the freezer." responds Alice. She shrugs at Kratchet and Donnelley -- no frank answers will be forthcoming now.

"There are only two in here," says Bandooma with disappointment.

"You'll have to share with Dalango." she replies, "The food shipment has been delayed."

"Grr... I hope they catch those bandits soon." says Bandooma, "I'm sure missing my ice creams."

Now Alice is really happy she hadn't answered Kratchet's question, but she does nod at the ladies in sympathy.

"Not shipping! How the fuck am I supposed to run this country if I can't get shipments!" Radamm throws his glass of wine into the fireplace in rage. He has just enough control that he doesn't toss it at a painting on the nearest wall or break apart the fancy chair he is sitting in.

"Who is holding up the shipment? Who specifically. Get that person on the line so I can talk with them face-to-face." he says.

On his desk screen comes the face of Alistair McCally, a VP of Turkistan Distributors. He looks as if he is not happy to get this call.

"How can I help you today, Radamm?" he says calmly but coolly.

"You can get that delayed food shipment rolling, Alistair. This is bullshit."

"Have you sent another payment? That's what the holdup is about."

"My people are starving and getting real angry, Alistair. Do you want to have the blood of food riots and a violent crackdown on your hands?"

"It's not in my hands anymore, Radamm, the International Payments Department is intervening. Your fastest solution is going to be scraping together a payment from somewhere else in your system and getting it sent, chop-chop."

Radamm gets a real sour look on his face, "Don't plan on taking any more junkets to our casino for a while, Alistair." He slams his fist on the cutoff key.

He stands and looks around. He says to Manilla, his cyber muse, "What have we got that can be used for a payment to Turkistan?"

Manilla is not surprised by this question, she responds immediately, "Not much left. I recommend diverting the bus driver's payroll."

Radamm thinks about this choice. He's about to give in to it when another idea hits him, a much better idea in his mind.

"We have enemies. It is now time to make them pay. And we can use what they pay to pay for what we need. Manilla, get me Demitri Stalink on the line."

The Revolutionary Police announce another parade. This one is surprising, and they are only giving twenty minute's notice on it. Alice dutifully closes her shop and dutifully heads for her spot on the main street.

This parade looks most unusual. It is quite short, only about twenty men. Leading the pack of Revolutionary Guards is Janz Kotch, another shopkeeper. They have him in front, pushing him along, hands over his head. Following behind the pack is a wagon carrying the contents of his shop.

"Traitor, traitor, traitor" chant the guards.

Lapcost, now a Revolutionary Guards leader, looks at the gathered shopkeepers who have come out. He looks at Alice in particular. He pauses the group and shouts out what is odd parade is all about.

"Janz Kotch, here, is a traitor to the country. He has stopped providing the people with the goods that they need to lead an honest and upright revolutionary life at honest and upright prices. He is no longer welcome here.

You other merchants! Take note! You keep raising your prices; you stop stocking your shelves, and your turn is coming!"

The "traitor" chanting begins again and the group marches on, headed for the central police station.

Radamm is in his office looking much happier. The traitor campaign is letting him pay more bills and distracting people from the slow progress of making the mine profitable. Things are far from perfect, but they are better.

He gets a knock on the door and in comes Lotran.

"This is a pleasant surprise, Ms. Lotran, you don't have an appointment."

"I'm afraid it's not going to be pleasant for you, Radamm. The kleptocrats are not happy with how things are progressing. Their businesses are suffering badly, and they are blaming you and your Traitor Campaign."

"How is my Traitor Campaign causing them to suffer? It is directed against shopkeepers in the Mining Sector."

"The international business community is taking note. And they don't like what they are seeing. Rather than pushing you they are pushing us. ...And it's working. We are taking action.

This is to let you know that, officially, you have become very sick, Radamm. You have decided you are now unable to fulfill the heavy demands of being President and are going to an undisclosed health spa to take cures."

Radamm looks at Lotran in amazement. Lotran makes a quick call on her communicator and two burly security types come in the office.

Lotran says, "I sorry we must make this so sudden, but it will smooth the transition considerably."

The security types escort Radamm out of the office.

Alice is at her altar praying again, and she is praying with deep fear and concern on her mind. The hologram of God is on the altar listening.

"Dear God! The Revolution Guards are sweeping through the sector. They are rounding up foreigners, like me, and sending them away. And they are taking our stuff! ... I don't have anything but what I have in this shop. I... I wasn't expecting anything like this. ...I am so scared. Lord, please help me!"

There is loud knocking on the front door. This is no neighbor looking for a cup of sugar.

Should she answer it? Should she hide? Should she run away and hide somewhere else?

The hologram of God tells her in a calm voice, "Answer the door, Alice."

With deep fear she rises, walks to the front door, unlocks it, and opens it.

At the door is Mrs. Kratchet, "Did you hear the latest? President Radamm has been removed. He's taken sick and is 'at some undisclosed location taking cures', as they are calling it."

Alice gasps, "Does this mean..."

Kratchet nods, "The Revolution Guard is being dismantled. No more raids in Mining Sector."

Alice hugs Kratchet in deep, deep relief.

<<<*>>>

Ms. Mandis is in Radamm's presidential office sitting at his desk. She is talking on the communicator screen with Kent Clark, the mining president Radamm kicked out, what now seems like ages ago.

"Yes, Mr. Clark, we are calling you back to run the mining company. We are now interested in running the mine in a modern way and making some serious money doing so. The nation needs it. Are you interested?"

Kent smiles widely, "It is a challenge I would love to take on, again, Lotran."

"You will not be alone in taking on new challenges. I'm bringing in Mr. Lee Ho-me, a technocrat from Singapore. Let me bring him on the line so you two can get acquainted."

Lotran makes a swipe and Lee Ho-me, a dignified looking forty-year-old of Chinese descent appears on the screen with the others. He begins talking.

"Good to meet you, Mr. Kent. This change planned for the mining operation is going to open a lot of opportunity doors. Now that we are..."

he puts his fingers up in quotes, "'doing it right', the cyber international financiers will get behind us. This means many things can happen quickly.

First, you will get a lot of resources to automate the mine, and thereby make it quite profitable quite quickly.

Second, the profits from your operation will be dedicated to making the lives of the Blabistanis better by transforming the mining sector into a state-of-the-art smart city. We will be rebuilding the city and reeducating the people. Thanks to your money and a state-of-art mix of human and cyber educating, we should be able to make Blabistani students top-of-the-line in education in two generations, and the first generation will be impressive, too.

That's what we have planned."

Lotran says, "Thank you, Mr. Lee." and to both of them, "Blabistan should now be able to take its place among the prosperous nations. This will be good news all around. The people will gain, the world will gain, and us kleptocrats will transform into just being filthy rich people." there are chuckles at this, "All-in-all, if things work out as *now* planned, this will be a great blessing all around. Santa Claus is coming to town... on steroids."

It's Time

Prelude

This story is about having babies and raising them in the 2050's.

It is about how all the interesting choices that new health technologies will bring change the choices a person faces once they decide to have a baby. It is also about a social difference -- raising the child as being part of a baby club, rather than being part of a marriage. I envision baby clubs as becoming a much more popular child raising institution in the 2050's -- as popular as marriage, or even more so.

The many choices in creating babies and raising them

Health technologies, artificial intelligence and powerful gene editing tools are going to be bringing about lots of changes in how children are created and raised. There are going to be many, many more choices in how to accomplish creating the zygote, incubating the fetus, and raising the child.

A woman or man who decides to have and raise a child is going to be facing all these choices and having to pick and choose between them. It's not going to be "birds and bees simple" anymore.

The origin of baby clubs

I foresee baby clubs as being a popular child raising lifestyle in the 2050's -- this is what the single mom and other alternative child raising environments of the 2010's will evolve into. It will become popular and respectable because it meshes more easily with the lifestyles of those who want to mix child raising with many activities other than marriage.

Marriage is going to continue evolving away from having child raising at its core. I envision marriage as getting wrapped up in Tattoos and T-Shirts issues -- it is going to become a lot more about personal expression and a lot less about getting children born and raised. This means marriage will evolve into a lifestyle choice that is a lot more expensive, a lot more elaborate and ritualized, and a lot more about being in love. It will be less about raising children and creating social ties between in-laws. This also means that while fewer people choose to indulge in it, those who do will be much more enthusiastic about their choice than average married people of the 20th century were. Some married folk will raise children, but many will be marrying for other reasons that center on other forms of personal expression such as the romance, the wedding and conducting social functions as a couple.

This evolution in what marriage is about will boost the popularity of baby clubs for those who want to have their child raising happen in simpler, less expensive, and more varied ways, and in an environment that is centered on child raising.

This story is exploring how all these choices will affect "baby making in the 2050's".

The Cast

In order of appearance --
Janalee Crosby -- 30's, AR engineer, into tennis and bicycling, she is ambitious and living an ambitious person lifestyle, not a necessity person lifestyle
Cynthia-975 -- Janalee's cyber muse, a moderately expensive one, who is physical and quite good at dealing with human personalities
Florence Manning -- 40's, personnel admin, also ambitious and who works at the same company as Janalee
Mildred Mingleforth -- 50's, the manager of the Hannaford Baby Club, she has a personable style
Robert Griswold -- 30's, clarinetist, modern dance dancer, slim, tall and good looking, Janalee has her eye on him for DNA

Cheri Albright -- 20's, attractive in a fashion model way, salesperson at Twinkle Star, an agency offering pro DNA that can be added to the zygote an aspiring parent is preparing to grow

Sly -- 20's, black and hoodie dressed, offering black market pro DNA

Nana Zima -- 30's, slim and black of South Sudanese descent, the director of the Nottingham baby club

Day in the life of

Janalee Crosby is a woman in her 30's and she takes care of herself.

She is slim and athletic and this morning she is out jogging in the park near her home. This is in the 2050's so the park is filled with beautiful lawns, bushes, flowers and trees maintained mostly by robotic gardeners. It also has its share of holosigns and as Janalee jogs by them they put up advertising customized to her needs and wants. Jogging is how she wakes herself up these days, her other weekly athletics are tennis matches, mountain bicycling and some swimming.

Jogging done, it is time to get ready for work and today is an office day. She showers, dresses, calls up a driverless car and heads to the office. On the way there she prepares for the meeting she will be conducting.

Janalee is a professional woman who works as an augmented reality (AR) engineer designing training courses for humans and cybers. Her current project is designing courses for assembling components that go into large cargo ships.

The meeting room at the office can hold about ten people around a central table. Today there are three others in the room with Janalee and four remote participants who are video linked in. Janalee opens the meeting.

She starts out brightly, "Hi all. Hope you all are feeling tip-top today." she waits for acknowledgements, they are mostly nods, all in this group are familiar with each other.

"And top of the morning to you, Janalee," adds Buster. He is over in Ireland. It is afternoon for him, but he is in the spirit of the meeting.

She continues, "OK. As I have messaged you things are going well. We are on schedule, but there have been some updates. The ship design is changing to incorporate this new tweak to the propulsion system that

Androdyne has come up with and we need to adjust the training to accommodate that."

Buster says, "They sent me some information on that, but it is outdated."

Janalee nods, "I just got a message on that as I was driving over here. It's an 'Oops!' and they will be sending updated info in three hours."

Buster is satisfied, "Good! Those bozos sure love to get ahead of themselves. This is isn't the first time they've done this."

Janalee says, "Yeah, but thank goodness they are getting quicker about acknowledging the boo-boos. I hate it when people say, 'Mistake... what mistake?' Those people are hard to work with."

There are nods all around to that, and the meeting goes on.

It is a busy day, time flies, but it ends in time for her to get home, change, and go over to the nearby dancing school for an evening class-and-dance session. There she is learning Irish step-dancing. She saw Riverdance and got inspired by the cadence the dancing produced -- it sounded neat. She and her group can't pound as hard and fast as the Riverdance folk, but they have a great time trying. They end the session with some wind-down ballroom dancing.

That finished, and feet pounded, she returns home. She winds down by reviewing some evening news and doing some VR gaming, then heads for bed.

It's time

It is morning the next day, and as she works through her bowl of breakfast cereal neither biking, dancing or designing are on her mind. Something much closer to home has her thoughts. As she finishes her cereal she opens up to Cynthia-975, her cyber muse.

"It's time."

"Time for what?"

"My baby."

"That's a surprise."

"Well, I've been thinking about it for a while. And now Kathy Kratch, my HR VP at work, has been saying at the pep talks that company policy has changed -- it is now that taking maternity leave every few years is a bright spot on our performance reviews. I guess management has jumped whole hog onto this 'Baby Gap' issue."

Cynthia takes this surprise in stride, "Well... you have watched the 'birds and bees' video courses, so you know what you're getting into.

The first big question is: What Baby Club do you want to join?

And the second one is: How natural do you want this baby to be?"

Janalee has been thinking about this for a while. Her answers are equally in stride, "I think I'll look first at getting into the one Florence is in. I like her, and she says great things about it."

"I'll check if they have an opening..." Cynthia gets on her internal communicator and gets a quick reply, "Yes, they do. Would you like me to schedule an interview?"

"Let me talk a bit with Florence first. We'll schedule while I'm at work."

"This means lots of change. You'll have to move out of here, you know."

"I'm ready for it."

"So what DNA do you want to use?"

Janalee doesn't have a quick answer to this one, "That one is tougher to answer. It's sure not something simple to decide."

"Do you want more advice?"

"...Yeah, I can use that."

"Who from? I can make recommendations."

"I hope so! Gosh! When this topic comes up it seems like half the world wants to talk about the best way to make babies and the other half about best ways to raise them. Does the baby club offer good advice on this?"

"They do, if you want sperm from an anonymous donor. The big question is: do you want DNA from someone specific? If so, that may take some more planning and negotiating."

Janalee is thinking hard about this. She has had boyfriends, and a couple would be great to "have in the baby's eyes" as the current slang goes. But there are other good possibilities, too. There are professional and celebrity sperm donors... and even wilder possibilities, like DNA from girls. She is still thinking hard about this one.

She tells Cynthia, "Yeah... no answer to that yet."

Enough for now. She gets up from the table, "Time for work. More talk when I get back."

Deep in thought she heads out the door. This time Cynthia has called up the driverless car for her. Once she is out the door Cynthia and the other home cyber take care of cleaning up after her.

At work Janalee gets into her engineering work, baby coming or not, there's lots to get done. When she takes a break she goes to the exercise room. This is also AR equipped. There she mixes some bicycle riding along scenic trails with a tennis match.

Then comes lunchtime at work and Janalee is eating with Florence Manning. Florence is a seasoned office worker in her 40's. She doesn't have Janalee's technical skills but she uses her wisdom about people to keep her position at the company. She keeps fit with yoga, modern dance and swimming.

Janalee says, "Florence, I'm thinking of joining you at the baby club."

Florence looks at her for a moment before commenting, "Starting early, are you?"

"I want to make sure this child gets raised right. I want a child who grows up to be a winner. If there are... unpleasant surprises... I want to have a chance to try again."

"Are you expecting unpleasant surprises?"

"No! I've checked my own DNA and it's fine. It's editable." then she adds, "But even these days life is full of surprises."

Florence takes a sip of coffee, "That part is still so true." she sighs, but doesn't bother to explain further.

"Well, we have 'new hopeful' introductions on Wednesday nights. Should we sign you up?"

"We should."

Both go into zombie mode for a few seconds to have their internal communicators handle the scheduling. Then they come back to real world interacting.

Florence gives her a pat on the arm, "You're on, dear. I'm looking forward to it."

Welcome to the club

It's Wednesday night. The Hannaford Baby Club is a nicely appointed twenty unit apartment complex that has been built specifically to support a baby club. Janalee has done video tours, now she is at the complex to do the in-person tour. Cynthia comes along as well -- she will be even more a part of the baby club environment than Janalee will.

The New Hopeful presentation explains the preliminaries. There are four other attendees besides Janalee, Florence and Cynthia. Mildred Mingleforth, the manager, is the presenter. Mildred is in her fifties, medium height and build, and has a personable style.

Mildred explains, "The apartments surround a courtyard that is a medley of playgrounds designed to support lots of styles of kids' activities and support kids of all ages up to those who are of entering high school age. The equipment in the playground is transformable. It can quickly be adapted between things such as merry-go-rounds and seesaws for elementary school kids to basketball courts and track courses for middle school kids.

Outside, on two of the sides, is a wooded creek flowing into a small pond. The kids can play in these. They are well monitored and equipped with rescue cyber.

Interspersed in between the apartments themselves are rooms that serve as activity centers. They can be used as labs and support other projects that the kids will get involved in such as dancing, yoga and plays. They can also be used for parties of various sorts."

After the presentation the hopefuls have interviews with club members. Mildred, Florence, Janalee and Cynthia meet in Mildred's office. They start right in.

"Any questions, Janalee?" asks Mildred.

Being here and putting so much attention into the baby business is getting Janalee a bit giggly-style excited -- kind of like she felt when she went on her first date. "Oh, dozens and dozens about this whole baby making and raising business, but just a couple about the club.

First, do you have any policies on mixing job and club activities?"

Mildred says, "Yes, we do." she looks at Florence, "Florence is quite well versed in them."

Florence says, "I do just fine with them. I can fill you in on those that affect our workplace."

Janalee is satisfied with that answer, "Second, I've still got a lot of questions about who's DNA to mix with mine. Can you help with that?"

Mildred smiles and says, "We've got six courses we offer on that, and we have regular guest lecturers. Plus, you can get lots and lots of advice from the other ladies. This is something they love talking about." she grins even more, "And, we've got plenty of examples to show off."

Janalee laughs at that, "OK. That's it for questions."

Mildred happily says, "OK. Let's go walk about."

The questions taken care of, they go on a tour. Much of the tour is about meeting other club members, with everyone watching how Janalee and Cynthia interact with club members and their kids. Janalee has done her homework. She and Cynthia began researching the club member resumes as soon as Janalee made her choice. And, likewise, the club members have seen Janalee's resume. As Janalee tours she and the club members greet each other much as they would acquaintances at a business convention.

The first person they meet is Angela Hotchkiss. She is in her sixties and a ten year member of the club. Her daughter, Alice, is nine. She is quite visibly sizing up both Janalee and Cynthia as she walks over to greet them.

"Hi Janalee, I'm Angela. Welcome to the club. What made you pick Hannaford?"

"Florence is my friend at work."

Angela smiles at that, "Excellent. We like connected members. Complete strangers are so unpredictable. And you have a good idea of what you're getting into. Good to meet you." and she walks on.

Next they meet Sally. She greets Janalee with a hug and asks, "Have you got a boyfriend these days? Is he likely to be visiting you here?"

"No, not at this point."

"Good. We've had some pretty strange ones show up once in a while. The fewer the better as far as I'm concerned." After a little more small talk she moves on too.

Next they walk by a couple of the playgrounds and then meet some kids. The adults Janalee has researched. The kids are brand new to her, and she is brand new to them.

In the playgrounds the kids are everywhere, and every age, and are up to all sorts of activities. Mixed in with them are cyber muses doing the tending -- they range from robot looking to very human looking, like Cynthia.

"My goodness, they certainly get involved in so many different activities," comments Janalee as they walk.

"We support variety here." replies Mildred, "Some accuse us of supporting Tiger Momism, but that's way beyond what we support."

"Tiger Momism?"

"That's when the kids are subjected to some really extreme environments -- the kinds that make them catch fevers and occasionally break arms and legs. We're not into that. We're much more loving."

Florence adds, "But we don't support Tender Snowflakism, either. That's when the kids never experience anything more violent than a mildly combative computer game or videos of some kittens whining for more milk from a mom."

Mildred adds, "The moms here want their kids to contribute meaningfully to society. The activities you see going on here are designed to encourage them to learn how to do that. These kids learn to aspire."

Florence says, "Speaking of kids, here comes mine." Five year old Patsy Manning comes running up to her mom and gives her a big hug.

"Hi mom, I just finished it," she says in an excited way.

"Finished what?"

Patsy motions for Lucy, her muse, to come over. Lucy comes over with a robot toy that looks like a pony.

Florence says, "Wow! How long have you been working on that, dear? It's so beautiful!"

Patsy is so proud, "For two days now. Lucy has been showing me how."

Florence admires it, then says, "Why don't you show it to Janalee, here. Janalee is my friend at work, and she may be coming here to live... if she likes it."

Patsy looks at her in a new light, and shows Janalee her work, "Hi Janalee." she says.

Janalee admires what she has made. "Wow! How old are you, dear?"

"I'm five."

Janalee is even more impressed, "I think I would have been six or seven before I could make something like that. You're learning a lot here, Patsy."

Patsy smiles at that. Lucy says, "We still have another project to work on this evening, Patsy, let's get back to that."

Patsy says, "Bye now." and she and Lucy head back to the lab room they came from.

Janalee has a thought, and says to Cynthia, "Are you ready to change your lifestyle, too?"

Cynthia nods, "Umm... I can see I'll be doing some learning, too."

As they are walking Janalee notices something new. She mentions it, "All the children I see are girls."

Mildred answers, "That's right. The club members are girl favoring."

"If I choose to go with a boy, will that make a difference?"

Mildred stops walking and faces her, "In a word, yes. A committee would have to vote on that choice, and, in truth, they won't support it without compelling reason."

Janalee is surprised to hear this, "Wow!"

Florence adds her editorial, "Boy children are so difficult, so noisy, so rambunctious. And when they grow up they become so useless, all men are useless, don'cha know."

Janalee listens and takes this in. This was a side of Florence she hadn't been aware of. The tour continues.

They meet Betsy Wollington, a woman in her sixties. She is cradling her six month old. "I planned ahead." she tells Janalee, "I had some eggs extracted and frozen when I was a twenty-year old." She smiles at the child. "This is the result."

"Where did you grow the fetus?" asks Janalee.

"In me. I had some uterus rejuvenating therapy."

Next they meet Lois Longmeer. She is in zombie mode as they walk up, and stays that way for about thirty seconds after they get close. Lois is in her early sixties and has an executive style to her manner. This is not surprising since she is the president of a medium-size company running a chain of urban vertical farms.

She pops out and begins the conversation, "Hi you're Janalee, right? Nice to meet you." she gives Janalee a quick hug, "Welcome to Hannaford. Are you enjoying it?"

Janalee nods, "This is quite an interesting place."

"Have you met my daughter, Anna? She's over there." Lois points, and waves at Anna to come over. Anna does. She is an eight year old. "Anna, this is Janalee. She is thinking about joining the club."

Anna gives Janalee a hug too. She says, "It's a really good place, Janalee, I really enjoy it. I have lots of friends here and I get good grades, too."

Janalee smiles at this testimony, "That sounds wonderful, Anna, it sounds like your mother has made some really good choices."

Anna nods her head enthusiastically, "I think so."

Lois tells her, "You can run along now, dear." and Anna does.

Lois turns to Janalee, "Anything else I can help you with, Janalee?"

"No..."

As she says that Lois goes back into zombie mode, she is a busy woman.

The meeting people is finished, the trio head for the entrance and a car is waiting when they get there.

Overall, the tour has surprised Janalee, she has learned a lot. As they walk for the entrance she comments, "There sure is a lot of variety in this club."

Mildred says, "We tolerate lots of variety in ages. But we try to select people with similar goals and aspirations. We want mothers who are comfortable being with each other."

At the entrance Mildred says, "Keep us posted on your thoughts, Janalee. We want to know if we should keep a space open." she smiles. Janalee smiles back at that. She takes one last look around and then she and Cynthia get in the car.

In the car as they are headed home Cynthia asks, "What do you think?"

"I think I'm going to have to make up my mind on this boy child issue real quickly."

Checking on some personal DNA

It is evening the next day. Janalee is at a community center. She is watching the performance of a small band using acoustic instruments. She is enjoying the show and in particular the performance of Robert Griswold, a tall slim good looking man in his 30's. He is also a clarinetist with the local symphony, into modern dance, and one of the guys that Janalee is thinking about mixing DNA with.

As the performance ends she applauds heartily. And as soon as Robert gets his equipment packed up they head out together for some dinner at a nice restaurant. They have known each other about nine months now and done some cuddling under the sheets, so this is a familiar dinner they are having -- not a lot of suspense involved.

Janalee pops the question, "Robert, I'm thinking of starting a child, and I'm thinking of mixing in your DNA. Just thinking, mind you. What are your thoughts?"

Robert is surprised to hear this, "Are we talking about mixing in wedding rings as well?"

"No. This is just DNA. I'll be living in a baby club and raising the child there."

"Well, my first thought is I'm honored. Thanks for making the offer." he eats a few more bites before he continues, "Do you want anything from me but DNA? Do you want support? Do you want help raising the child?"

"No. I'll be supporting the child myself, and I and the baby club will be doing the raising."

"What happens if I really want to get involved in the raising?"

Now it's Janalee's turn to be surprised, "Do you? If you do, I'll have to take it up with the baby club. It could be an issue, many are pretty woman-centric."

Robert looks down at his hands, "This is something I've been giving some thought to, as well. The thought of just handing off sperm and waving bye-bye rubs me the wrong way." He looks up at Janalee, "It may not be stereotypical guy thinking, but it's how I think."

"So, your sperm comes stringy... I mean, with strings?"

Robert grins, "And it might be sticky too."

They both laugh at this, and then the conversation moves on to other topics. This is a pleasant night out, not a decision making night out.

Back at home Janalee does some more online research about commercial DNA sources. There are a lot, and they do lots of advertising. To her it feels like what her grandma encountered when she decided to buy a new car -- lots of images of success and power mixed in with lots of descriptions of features. Cynthia is there helping her.

The advertisement that grabs her heart-strings is the one that shows a young man graduating from college. He hugs his mom for a moment then switches to just holding hands and says, "Thanks mom. I know you did a lot of sacrificing for me. But... I think it's paying off. And... I really love you!" he hugs her again, for longer this time, and she melts into it. While they are hugging this time the ad streams by a list of accomplishments the boy has already achieved. When they break apart this time the boy gives his mom a mischievous grin and says, "Would you like to meet my girlfriend?" He points, and she is a lovely looking coed standing nearby also in cap and gown.

Her heart loves it, but her analytic side isn't as impressed, there are several other brands touting better features. On the features presented side she likes one from Twinkle Star DNA Agency.

And overriding all the talk about features is the expense -- the really attractive sperm packages don't come cheaply. Thinking back on her grandma again, the top-of-the-line sperm prices seemed comparable to what she would have spent on a car.

She mutters to Cynthia, "It looks like there's going to be some serious budget planning happening along with this baby."

"I'm ready to help."

For the next couple hours Janalee and Cynthia work up spreadsheets about what a baby is going to cost over the years. As an option, they also work up what two babies will cost to see what kinds of savings and synergies come with two.

Overall, it's complicated. There are so many ways to spend money on the child, and there are so many ways the government will offer help, but with so many strings attached! Even with Cynthia's help it takes most of the evening to sort through just the basic options. By the end of the evening Janalee has a grasp of the basic finances, but these will vary a lot based on the choices she makes for DNA, incubating, baby club and how many kids she decides to have. Whew!

And she is now back to thinking about the boy/girl choice. It's such a simple one, but it affects so many other elements, such as what clubs she can get into. And then…

"My Goodness!" she throws her hands up in dismay. She takes a break from all this "Yes, but-ing" (as she calls it) and goes back to thinking about projects at work. Compared to what she has just been through, these are much easier to comprehend.

Inside/outside me?

Back at work, and on a lunch break, Janalee asks Florence, "Did you grow Patsy inside you?"

"No. The doctors told me my uterus was too stiff. I could have done some therapy, but I decided I'd let an incubator take care of Patsy while I took care of my dance technique. I was just mastering some serious new moves at the time. It paid off: Once I got those mastered I transferred to the New Conceptions Dance company."

Janalee is impressed hearing that, "Congrats on that!"

"So it turned out well. Patsy has come out well and my aspirations in both children and dance have been comfortably accomplished. Speaking of, how's your progress coming?"

"So much to think about! And this boy issue keeps coming back into my mind." she looks at Florence, "I haven't decided, but if I go boy I'll likely be looking for a different club."

Florence gives her arm an understanding pat, "I understand, and if you want a recommendation, let me know."

Janalee is relieved at that. Lunch continues with different topics coming to the fore.

Checking out Pro DNA

Janalee is in the office of Twinkle Star DNA Agency. It is a professional person DNA offering agency. This is DNA from people who make some money offering theirs. Some make a pretty good living off of it, and some offer it at a price that fits in her budget.

Cheri Albright, the lady she talks to, is pleasant in a salesperson selling high end product sort of way. She is in her twenties and has worked on her looks like she is a fashion model. When Janalee comes into her office she waves and says cheerily, "Hi," she goes into zombie mode for just a second, "Janalee, what can I help you with today?"

"I'm exploring getting some DNA."

"For you?"

"Yes, for me. It's time."

"Good. I like dealing with the person directly, and someone who is ready. We get so many curious relatives and friends coming in. They should be checking this information out online."

"So, I've been looking at James Stewart DNA. It seems to have the features I want."

"Do you want it straight, or with modifications?"

"Oh, the basic ones. Things like Vitamin C metabolism."

"OK. The contract for basic is quite simple. But the more modifications you want the more complex it becomes. Those modifications come from third parties working over the DNA."

"You don't do those in-house?"

"We package, we don't create. Creating is the work of talent. We are a talent agency, not the talent."

Janalee talks a bit longer. She discovers that, yes, this using pro DNA gets legally complicated as well as expensive. "But if it makes a better child..." is Cheri's theme as they talk. The problem here is that Janalee has read lots of articles by critics who say this pro DNA business is just a racket that feeds on hopes and dreams. That explains the contract hoop-jumping that comes with this choice, it is designed to protect the talent and the agency from remorse.

By the time they are finished, her chat with Cheri has not reassured her. She sighs as she is walking out through the lobby. More research ahead.

The Black Market

Janalee really does want the best for this upcoming child.

This is a field that is changing rapidly and one that picked up a lot of FDA heritage -- which means, as she discovered at Twinkle Star, lots of legal involvement. But there are alternatives that are not so legally entwined. There are players who picked up their heritage from offering mind altering, not FDA approved offerings. Their mentors were in the cannabis and designer drug scene, and they play a lot faster, looser and cheaper.

She is now thinking hard about exploring something in this arena. The first thing she discovers is that these people aren't nearly as straightforward to contact. But she finally locates one with an office that is currently behind an abortion clinic. She heads over.

The person she talks with here is night-and-day different from Cheri Albright. It is a man dressed in a hoodie who looks like a gangsta-rappa wannabe.

"Call me Sly." he says when she comes in, "What'cha here for?"

"DNA. Call me Toots," responds Janalee, getting into the mood.

"OK, Toots, how much do you want to spend?"

"What are you offering for a kilobuck?"

Sly makes a face at that, "That's all, Toots? I guess you're looking for a bargain." He sighs, gets up and looks through a filing cabinet behind his desk -- a for-real filing cabinet with for-real paper in it -- and pulls out a single page brochure.

"This doesn't leave the room, and no pictures!" he says as he hands it to her. The page has a picture of a handsome man at the top and specifications below it.

Janalee looks it over, the specs are impressive for the price. She looks up and says, "If I go for this, what happens?"

"You get me an egg, or three, and two weeks later I give you a zygote, or two, or three. What you do with them after I give them to you is purely up to you. But, if they grow up, they should be some red-hot items." he looks her over, "In your case, toots, white-hot." and grins at that.

"And if they aren't? If they come out DOA? If they come out... defective?"

"If they're DOA, we try, try again. If they're defective... ah well, try, try again, on your dollar." then adds, "These people I work with are top-of-the-line. Sorry, I can't personally give you a list of recommenders, but check on the street. My rep is rock-solid, and my people do primo work."

Janalee has the message. She gets up and says, "OK. I'll do some checking. I'll be back in touch if I like what I find out."

"Don't take any wooden nickels." he is back into zombie mode before she even turns around to head out.

Who to talk to now?

It's breakfast time a week later. Work has been keeping Janalee busy so baby issues have waited. She doesn't feel bad about that, there has been so much to think about. But this morning Cynthia brings it up again.

She has a warm glow about her when she says, "Janalee, I think I may have found the right club for you."

"Really?" Janalee perks up and pauses gobbling.

"I've been researching. There is a club that's near your work that has women who are both ambitious and tolerant of boys. The biggest catch is that most of the women are immigrants. This is why they are both tolerant of women with many different aspirations, and support raising aspiring children. The big question I can't answer is: are these women too diverse for you to be comfortable living with?"

"Well... schedule a visit and we'll find out." Janalee resumes eating with more enthusiasm. Ambitious women don't sound like a problem to her.

Nottingham Club

Janalee tours the Nottingham Baby Club. This is the club Cynthia has discovered. Physically, it looks a lot like the Hannaford club, but the members look quite different.

Janalee admits it to herself, she is surprised at the diversity she sees as she walks the grounds. There are women who are white and dressed as conventionally as she is, but they are in the minority. Surrounding them in large numbers are African and Asian women who are dressed in many different ways -- some are fully covered, likely following religious dictates, some with fashion styles that are just different, and some with more revealing styles than are conventional out on the streets. Not only are the fashions mixed, the conversations she hears are taking place in different languages as well. All in all, quite a mix.

The children are more consistent. They are all doing lots of studying and free-style recess playing, and spend little time in zombie mode.

Nana Zima, the director here, and Janalee's host for this tour explains, "Aspiration is the important goal here, and we support diversity because we feel that supports aspiration. A child who experiences many ways of doing things is learning to adapt and solve problems in creative ways. They are also learning tolerance, so they can use these skills they learn in many different situations. They can be surprised by their surroundings and still function -- but it is also harder to surprise them.

In addition to the learning you see going on here, we also encourage a lot of travel. We encourage mothers to take their children on vacations to exotic places when the children are young, and encourage them to get their children involved in exchange programs when they get into formal education programs."

Janalee asks, "What happens if the kids start fighting?"

"The mothers break it up. Those that are closest take action. They don't worry about the kid's cultural backgrounds, we are all mothers to all the kids here."

"What if the mother teaches someone else's child their cultural ways?"

"Oh, they do! All the time. The child learns and learns which cultural ways are their own, and which are different. But, having learned, they then get to pick and choose. That is part of tolerance." She points, "See young Achmed over there?" the child she is pointing at is wearing a turban and a gaudy American-style T-shirt. "We have some mix-and-matching happening, and that's OK. The only time they have to be straight about their fashions is when they are attending church services, and they learn that. They also learn by attending church services other than their own as guests."

Janalee thinks a bit about what she sees and then says, "That's a lot of learning."

"It is. It is part of the maximum aspiring that we offer our children."

She notices a lot more men in the mix than she saw at Hannaford. She asks, "How do you treat men?"

"A good question. We welcome them... as long as they aren't being parasites. We have no interest in 'living in basement types'. The ones you see here are boyfriends. But some are more special than that." she points to the building next door, "Next door to our club is an associate male-oriented baby club, Nottingham North. Many of the men you see here are raising children there."

They walk and talk for a bit longer, then the tour ends and Janalee heads home. Janalee has been favorably impressed by what she has seen.

Choices are made

Janalee is at dinner with Robert again. She has come up with a plan and she presents it to him.

"Robert, I like you, and after the researching I have done, I still like your DNA." Robert grins at that, "And I have come up with a plan.

There is a baby club in town, Nottingham, that supports diversity and achievement. It looks like a good place. Just as important, next to it is a male-oriented baby club, Nottingham North, and the members of the two clubs mix a lot. Check it out."

Janalee waits while Robert goes into zombie mode and does some quick research on Nottingham North. He comes out and nods at her.

"You wanted to be involved with your DNA. I'm thinking you can get seriously involved. I'll make a couple of children, and you can raise your's at Nottingham North. We can mix and match.

What do you think?"

Robert thinks a bit, not too long, then leans over the table and gives her a quick kiss.

"I think this is why I really like being your friend, Janalee. This sounds great."

Epilog

This story is an example of how child raising will be quite different in the 2050's. There will be dramatic technology differences and dramatic social differences. It's not going to be grandma's style of baby making and child raising.

Megan's Party Time

Introduction

This is another tale about having a night on the town. This time it is being celebrated by someone living the necessity lifestyle, not the ambitious lifestyle.

The cast

Megan Silver/(Lola Lovegood in AR) -- 20, is deep in the necessity lifestyle, she lives on necessity money, fashion is her deep interest, her athletic interest is dancing in AR and real world
Hilda-956 -- Megan's cyber muse, Hilda is basic and purely virtual, she is what necessity money provides
Annabell Deloran -- 20, Megan's real-world friend, also a necessity lifestyler, and also into fashion and looking good
Anton Miller/(Mickey Moore in AR) -- 40, a shy, somewhat ambitious type in real life and an AR character in Megan's game
Pepper Corn -- 30, an AR reality celebrity who is also a real person, and famous for hyper-dedicated promoting style

Success and celebrating

Megan Silver is in her room in her parent's apartment. She has spent the last five hours working through a project in RunWayWild, an AR fashion model game. In the game she plays Lola Lovegood and she designs clothes then models them herself on the runway at fancy fashion shows. In real life she is a slim, good looking twenty-year-old woman, and in the AR game she is a stunning, fully fashion-enhanced version of her real-life self.

So stunning that Mickey Moore, her boss in the game, tells her, "Megan, that was some fine work today. It was so good that at the party tonight I'm introducing you to Pepper Corn."

In game Megan gasps and says, "Really?"

Mickey nods in a significant way and adds, "It could lead to some promo work for you." Underneath him, not part of the scene, a message flashes "LEVEL UP".

Megan is delighted. She has won this session big time. She pops out of the game, gives herself a nice satisfying huff, then, using her internal communicator, talks with Hilda-956, her cyber muse. Hilda is purely virtual and hasn't been upgraded for a least a year. But that's all Megan's necessity money allowance can buy her for now.

"Hilda, I've leveled up again!" she announces proudly, "Time to go out and do some real world celebrating... and exercising."

Hilda says, "Good. Going out will let the real world see your creations."

Megan smiles at that. She likes it when she can flaunt her AR stuff outside. She likes it... but Eww... those boys and their catcalls! But then again, when she doesn't get those she knows she is off track on her designing.

She gives her human friend Annabell a call on her communicator, "Meet you at Garfunkle's?" she texts. In a few seconds Annabell texts back, "I'm headed over now."

Megan heads out the door.

In the real world Megan is dressed in physical clothes as nicely as her necessity money allowance will allow -- which are nice in a standard way, but they sure can't match luxury money looks. But in addition she also has her AR looks, and the AR look is much, much better. She wears her AR fashions outside, so those walking by who look at her in AR see something much flashier and more fashionable than those who look at the real world her. As she journeys outside, Megan watches those who are watching her, and she can sure tell who is using which view system.

At Garfunkle's she meets up with her friend Annabell Deloran, another twenty-year-old who is also a necessity lifestyler, and also into fashion and looking good. One big difference in their looks is that Annabell likes her tattoos on her real-world body while Megan prefers her's in AR -- she likes being able to change them from month to month. Both like them flashy and sparkly, literally.

Once inside, both head for the necessity buffet table and stock up. Both come back to their table with only a half plate-full. Both are into watching their weight and neither can afford wearables that can compensate for serious overeating -- those are a new hot item and getting them takes some serious luxury money. There are protesters who want such wearables defined as rights items -- items which necessity money can buy -- but it's likely to be next year before progress is made on that front.

"Party after this?" asks Annabell as they munch away.

"Ummm," says Megan. She wants to flaunt her new AR looks. She wants to see if real-world humans appreciate what she has come up with as much as her AR game does. And party time is good vigorous exercise like she promised Hilda. They finish up and head for RetroDisco, just down the block.

There is a line to get in, but it's a slow night so it looks like only a ten minute wait and they get in line. While they are waiting they join the rest of the line in whiling away their time in zombie mode on their communicators. They get to the front of the line and the bouncer -- robotic tonight -- gives them a quick once-over look and lets them in as another group leaves. Inside, the RetroDisco is dark and noisy with lots of lights flashing in the stage and dance areas.

They see friends Rhonda and Alice already at a table with empty chairs. They join them, and that done Megan heads straight for the dance floor. She is feeling hot tonight and ready to show off her style. Here she minds not at all when the boys ogle her -- that's what she's here for. At home she spends lots of her break time between AR fashion projects working on AR dancing routines. Here she gets to put what she's been practicing into real-world action.

There's lots of competition on the dance floor. But she looks hot enough that after five minutes of getting into the music she is joined by Jimmy, a dapper looking boy she has danced with before who also has good moves. They hit it off for a couple more tunes, then take a break. Megan invites him over to the table and he comes.

Before she sits Megan says, "Whew! I'm getting a drink, first." She heads for the necessity bar. The necessity bar is a vending machine with a dozen or so choices -- soft drinks, fruit drinks, one beer selection and a flavored vodka selection -- the vodka flavors are the other drinks. (The luxury bar across the room has bartenders with a wide selection of pricey brand names behind them, and barmaids.)

Some people like to get alcohol high and they get into the vodka and beer. Megan prefers a wearable high because it's a lot quicker to turn on and off and there is less hangover. She will be getting a juice.

As she gets in line for the vending machine a man sitting at a table on the other side of the dance floor gets up and walks over to her. Megan watches him in AR. He looks to be in his 40's and is dressed and groomed like he is a Hollywood producer -- it is likely his form of cosplay. As he walks toward Megan he looks like he likes what he sees and when he gets close he says, "Nice dancing, sweetheart, I'm impressed. Would you like to get a *real drink* and come sit with me a while?" He says this in a nicely polished voice. He's doing his cosplay well, too.

"No thanks," she says in a brush-off way and turns from him to watch the line. He gives her a couple seconds to change her mind, then shrugs, and walks back to his table.

When her turn comes she gets a diet fruit drink and heads back to her table.

After she sits down and downs half her drink, she dials her wearable from "Dance Floor" mode into "Party Whoopee" mode. The wearable will take a few minutes to complete the transition. She takes another swallow and then relaxes and looks around the table. She sees that two other boys have joined them. Seven people -- it's nice, and her wearable is warming her up nicely too so she's not feeling shy, or brush-offy anymore.

The conversation at the table turns to sports -- nothing she is interested in. Her eyes wander. She sees that man, still at his table, still alone. She switches her view from AR to real world. His face is not quite as striking looking, but he is still good looking, and he is still dressed in a suit, but not nearly as expensive looking.

"Hmm..." thinks Megan, "Yeah, he could be a for-real Sugar Daddy."

Now that her whoopee mode thinking has kicked in she is feeling more daring. She makes a choice.

"Excuse me a minute, folks," she tells the table crowd, and she walks over to the man at his table. She is feeling like she wants to be watched, she switches back to looking at the world through AR, and she walks over in a way that shows her runway style.

"Still OK if I sit with you?" she asks in a sultry way. The man looks happy at both her walk and her choice and happily motions her into a chair.

"That was some nice dancing you did earlier, and I love your dress. In fact, I think I've seen it before."

This surprises her since she only finished making it hours ago. She asks suspiciously, "Really... where did you see it?"

"In RunWayWild," he answers calmly.

Megan is surprised to hear that, "You play?"

He nods, "I'm Mickey Moore," and then grins.

Megan's surprise doubles, "Mickey! I thought Mickey was NPC!" referring to non-player characters in the game.

He grins at that, "Much of the time I am, the real world keeps me busy so my in-game muse runs me. But I wasn't this afternoon." he grins even wider, "And I can't tell you how surprised I am to see you here... Lola."

He reaches across to fist bump, "Anton Miller in real life. You're..."

She bumps back, "Megan Silver," she is still amazed at the coincidence but finally thinks of something to say, "Would you like to join us at our table?" she points back to where her friends are and gets really happy. She stands up and gently pulls on one of his arms, "Come on."

Now it is his turn to be pleasantly surprised. He gets up and they walk over together.

Back at the table Megan introduces Anton, "This is a guy I know from RunWayWild. Mickey... er Anton meet Annabell, Rhonda, Alice, Billy and a couple other friends." She has forgotten the other boy's names.

"Alex and Sam" they introduce themselves.

"Anton plays Mickey Moore in RunWayWild. He's my producer." Megan finishes explaining.

"Nice to meet you all." Anton just sits, no fist bumping this time. Everyone goes into zombie mode for a couple seconds of resume consulting on the newcomer, then gets back into party-mode chatting. That goes on for only a minute, or so, before Annabell gets restless.

"Our turn now," says Annabell, and she and Alex get up and head for the dance floor.

The others watch the pair for just one dance. They are good, but now the rest are all ready to get into active mode too. The table empties. Megan dances with Anton.

With the help of their wearables everyone gets into dancing mood quickly and easily. The dancing is vigorous and lots of fun for all. Anton and Megan get into a tango-style sequence and she gets to do wild dips in his strong arms. She is loving it, he is loving it. When they come back up after the third dip she gives him a quick kiss on the lips. After three songs everyone has had their share for now and they head back to the table.

Megan is thinking hard about a question she thinks will come up, "If Anton wants to take me home, do I let him?" Anton is looking at her and smiling. She is pretty sure the question will come. She goes into zombie mode for a moment and consults with Hilda. "What do you think, Hilda, should I go home with him?"

Hilda replies, "I've checked his background. He doesn't have any felonies and he's not on any social shaming lists. Those are pluses. On the negative side he is not on any dating sites, so there are no reviews of his dating abilities."

Not long after Megan's checking Anton announces, "OK. That's it for me this evening. I've had a fun time all. Megan, thanks for the invite, and I'll see you in-game later." Anton gets up and walks away heading out.

The rest are a bit surprised at that. Annabell asks Megan, "He didn't invite you to come with him?"

"Umm... It's our first time meeting in person." It's the best excuse she can come up with.

Annabell thinks for a moment then says, "Ah well, not a hook-up guy, I guess. ...I wonder why he comes here?"

"I think he's all into looking." says Alex. "I've seen him here a few times."

The group continues their partying and breaks up an hour later. Alex is the first. He dials down his wearables to normal and says, "Had a great time. I'm outta here." Megan and the rest follow suit. She and Annabell walk out and walk home.

As they walk down the street Annabell asks, "Think anything will happen with this Anton-guy?"

Megan shrugs, "I'll find out more in-game."

Annabell puts her hand on Megan's arm, "What to spend the night at my place?"

Megan stops and looks at her, "Some VR adventuring?"

"I was thinking of some *PR* adventuring -- physical reality, slumber party plus." She keeps looking at Megan.

Megan looks back at her for a few seconds, shrugs, then gives a flirty smile back, "Sounds like fun to me."

"Good! We'll sneak in quietly so Mom and Dad don't wake up."

Busy day

It's 9AM before Megan finally makes it back home.

"Some breakfast, dear?" asks her mother as Megan heads for her room.

"I had it at Annabell's," she says quietly and without pausing.

When she gets to her room she pulls out a snack and a drink, then slumps into her chair and fires up her computer -- it's been a long day and night.

But today is going to be just as busy. There's a protest to attend in the real world, then more designing to be done in RunWayWild, and if Mickey Moore has been successful she has that Corn meet-up coming up.

The protest is five blocks away in a town square. Megan walks there carrying her holoplacard, and little else -- she has been warned that these things can get rowdy and she doesn't want to lose stuff to hooligans. It's not so much that they steal stuff, if they do that they get spotted on the surveillance and picked up, but they will trash stuff, including clothing. They will rip and tear, and toss paint, and even ickier stuff.

She is protesting for free-range kale to become a right, a necessity food. This is something really meaningful to her! And this is a bit unusual, too. Most of the time she is out here protesting things that to her heart sound like good causes, but are happening in distant parts of the world -- like saving polar bears. But the result of this being a "down home issue" is that the counter protesters are more numerous and vocal than usual. In this session they get really ugly and insulting.

She hears shouts of "Can't kale something, sweetie?" and "Go back in your basement!" and some really old codger yells, "Get a job!" What a weirdo! By the end of the session her ears are certainly burning in a figurative way.

When she finishes and gets back to designing in RunWayWild there is a lot of red and shiny silk in the styles she creates this session.

She doesn't meet up with Mickey Moore until six hours later. He comes up with a big smile on his face.

"Success! I've arranged an audition with a Pepper clone. This is a trial, if it goes well, then you'll move up the line and get a big step closer to what will become something high-profile and public."

Megan lets out a "Whoopee!" and gives him a big hug and a light kiss. Then she pops a question that is even higher priority in her thinking of the moment, "Did you have a good time at the club last night?"

Mickey is surprised by the question, he's been doing just in-game thinking. It takes a moment to adjust but then he answers, "Why, yes, I had a great time there."

"Do you want to meet up again... tonight?"

He smiles, "I'd be delighted. But now, let's get back to the in-game issues. There's a lot of preparing we... you... need to do." They both get back into full in-game thinking about this upcoming audition.

Back at the club that night Anton seems a little distant to Megan. She knows him now, and likes what she knows, so she is up for some cuddling, but he is just into some more dancing and talking.

She finally asks, "Do you have a girlfriend?"

Anton looks at her before he answers, "What difference would that make?"

"You're acting like you're married or something."

Anton laughs when he hears that, "No. I'm not married, and I don't have a real world girlfriend. I do my girlfriending in-game."

"Why is that?"

"Too many complications can pop up in real-world romance. I've heard, and seen, too many horror stories."

"Humph! I thought just women got broken hearts."

"Women get broken hearts, men get broken wallets. Both are ugly. It's just not worth it."

"I'm curious, has it happened to you?"

Anton shakes his head, "No. I've been careful."

Megan nods, "It hasn't happened to me, either."

She smiles at him, and Anton smiles back. Megan thinks, "It's a warm smile, some ice might be melting." They finish the night with some more dance and talk, and then go their separate ways again.

First Corn fashion audition

Lola Lovegood is in front of Pepper-375, a Pepper Corn cyber who imitates Pepper's AR presence. It looks for all the world like Pepper and acts like her too. The big difference is Lola is one of dozens of girls getting interviewed all at the same time by half a dozen cybers, while the real Pepper is elsewhere recording one of her shows.

As Lola walks in Pepper starts right up, "OK, dear, show me your stuff. Show me why I should be paying extra attention to you, and not some other wannabe."

Lola first fashion walks around in front of her in a sultry way, tattoos flashing underneath her gown like she is a Christmas tree high on mushrooms, and the gown's smart fabric is letting the tattoos show through. It's quite a sight, and then Lola walks right up to Pepper and says, "See this material. It's revealing, isn't it? It shows the skin underneath but shows it as if it's a fifteen yearold's skin... smooth, flawless and sooo flexible. You're not going to find this fabric on some other wannabe."

Pepper is impressed, a little impressed, she says, "Umm yes, good stuff... Good start! Any more?"

Lola on-the-spot changes her outfit into one that is form-fitting and goes into a vigorous dance routine with the tattoos still sparkling wildly. She looks good doing it and the Pepper cyber is more impressed -- enough that she says, "Yes... yes... Looking good, deary! Let's schedule you for a second level audition."

This is just what Lola wanted to hear. She blows Pepper a kiss, "Thank you! Thank you!" and manages to walk calmly out. As soon as the door closes she changes her outfit again to something street suitable, turns off the tattoos, rushes downstairs into the lobby, and there gives Mickey Moore a big, big hug. "I made it! I'm headed for second level," she says in a joyful way.

Mickey hugs her back and says, "Hot dog! I knew you could do it! Another giant leap on the road to BigTimesVille!" He admires her for a couple seconds and then gives her a kiss on the lips. The kiss lingers, she melts in his arms, and in a couple seconds her tattoos turn on again and sparkle even more wildly. They both laugh at that. She turns them off again and they head out, arm in arm, to a congratulatory drink at the bar next door.

And once again, Megan gets a level up.

Family time

It is evening. Megan is in her room in front of her main computer screen. She is thinking about heading out when she gets a text from her mother on her communicator, *You haven't joined us for a meal for so long, dear, come down to dinner tonight, will you?*

She thinks for a few seconds, then responds, *OK*. She dresses for family and heads down to the dinner table. Mom and Dad are both there. The meal at her place is something that came out of the freezer, just like what her parents have in front of them.

"So, what's keeping you busy these days?" asks Dad.

"Oh, my RunWayWild game is filling my time here. I'm at level 34 in that, and I've been doing some really neat designing in that. Would you like to see?"

"Sure."

"Pop on your goggles."

He does, and Megan brings up one of her dress designs. She brings up a summer dress for him, not one the hot items like she was showing Pepper.

"Looks good," he says.

"They are good enough that I'm showcasing them for Pepper Corn," she says proudly.

"Really?" says mom, "That sounds like quite an accomplishment."

Megan nods.

"And I had a surprise a couple days ago. It turns out one of the people I play with in-game lives in the neighborhood. We met a couple nights ago at RetroDisco."

"You go there?" says Dad.

Mom says, "Of course, she's a big girl now."

Dad thinks a moment, smiles, and says, "Yeah, I guess so."

"That's where I get my exercise." she explains, "I'm getting real good at dancing as well as designing. That helped me get the Pepper showcase."

"Who is this neighborhood person you met?" Mom asks.

"Oh... just someone," Megan hedges, then asks, "What's been keeping you busy, Dad?"

"Oh, the gardening is going well. That punk teenager next to me has given up his plot."

"Harry!" Mom interrupts, "We're at dinner!"

Dad looks just a little sheepish, "Well, he *is* a punk. He didn't keep his plot properly weeded, and those kept spreading my way. Anyway, he decided to go with more music instead -- joined a band that goes on lots of gigs. I talked to the room supervisor and I'm taking over his plot... for a season at least... and longer if no one else wants the space."

"That will keep you busy," says Megan.

"Indeed, and let me bring in those bing cherry plants I've been wanting to grow."

Mom asks, "Don't those grow on trees?"

"Yeah! Like money does." Dad grins, "The all-natural ones do. I'm getting a new style that grows on bushes. And they grow fast, too, only three months before they are ready to pollinate."

And the conversation goes on.

More party time

Megan is back at RetroDisco. She is looking for Anton but doesn't see him. In between looking she is doing lots of dancing, the success of the Pepper interview is inspiring her and she wants to get even better.

Finally, she sits at the table, recovering from dancing, and gives him a text, *Coming over tonight?*

The answer comes back quickly, *Can't make it, real-world duty calls. Catcha later.*

This isn't what she wanted to hear, so she looks around. "Stay and play, or head home?" she is thinking.

She decides to stay and play, but shortly after making this choice she gets a text from Annabell, *Want to come over?*

She texts back, *Yeah! I can do that.*

So much for staying and playing, she heads over for some more slumber party fun. It may not be Anton but it is fun.

More Corn success

Lola finishes another runway walk for Pepper-375 on a private stage next to her office. Pepper smiles, gives a big sigh of anticipation, and says, "I'm recommending you for an audition with the real Pepper."

"Really!" says Lola, and Pepper nods. Pepper goes into zombie mode to talk with the real Pepper, and on Megan's screen there is another LEVEL UP flashing. "It's on." Pepper says as she comes out of zombie mode, "I'll send your calendar the time and place."

As Lola leaves the room Pepper advises her, "You have a week to prepare. Bring your best, your very best."

Lola races down the corridor, and hugs Mickey and swirls him around when she gets to the lobby. "I'm on." she says, "Real Pepper audition next week." Now it's Mickey's turn to swirl Lola around. He plants a big kiss on her neck, "Incredible... and wonderful! ...all mixed together!"

"Will you be at Retro tonight. This calls for some real-world celebrating too." Mickey smiles and plants another kiss, "Let's see if we can level you up there, too."

At the Retro Anton and Megan are the center of attraction at the table Megan and her friends are sitting around. They dance spectacularly and smirk and snuggle at the table.

They do this for about an hour. But then Anton goes into zombie mode and when he comes out he dials down his wearables a bit.

He's soberer when he says to Megan, "Hey, I just got a call, and something's come up. I'm going to be traveling next week, in real world. I'm going to Timbuktu, of all places. And I'll be there for a month."

"You mean, in Africa?" says Sam. Anton nods, "It's work-related and it's going to be intense." he looks at Megan, "Which means I'll be NPC in Runway for the next month."

He takes her hand, "But, I can use an assistant there, in Timbuktu, if you're interested."

"I am! I am!" announces Sam. Everyone laughs.

Megan thinks, but only for a few seconds, "I would be leaving my friends, and RunWayWild, right?"

"You would."

She shakes her head, "Thanks. Really, thanks. But I really enjoy being with my friends. And I'm really looking forward to that audition with Pepper. Thanks, but I'll pass."

Anton looks at her for a few seconds as well, "You're sure?" she nods, he says, "Ah well."

He looks around, then dials up his wearables again. "OK. Well, let's make the best of tonight while we can."

And they all do.

Life goes on

A week passes. Megan works hard on another design and Lola goes to the audition. She levels up again there, and real Pepper says she is most impressed. But by two weeks later Lola still hasn't been invited to anything further. She wonders if Anton had been running Mickey if that would have tipped the scale some more in her favor.

Anton in real world has been sending texts, but just once in a while. He says he's really busy, and from how brief the texts are, it seems this is really so.

So for Megan life goes on... in RunWayWild and at Retrodisco. She keeps working hard on her fashion and dancing and keeps having fun with her friends.

Time flies, and she is getting what she wants out of life.

Epilog

This is a story of living life in the necessity lifestyle. This is a lifestyle that keeps a person happy and engaged, even in they aren't shaking up the world of 2050.

Epilog: What's coming after the 2050's?

Introduction

I have written a lot about the changes in technology and society that will be coming to humanity between the 2010's and the 2050's. The changes will be world shaking.

But as I envision the 2050's being world shaking, I'm also a bit unsettled. *I'm unsettled because I can't envision much change coming after the 2050's.*

Back to monotony

I envision the constant change that humanity has been experiencing since the start of the Industrial Revolution in the 1700's coming to an abrupt end in the 2050's. This ending of change will happen for the following reasons:

- *The Cyber Age will begin* -- the post 2050's will be a time when cyber becomes self-aware and will then dramatically surpass humanity in intelligence. It will also become much larger in quantity, it will be mostly inhabiting cyberspace but there will be a lot of that space.

The result of these trends is that cyber will be running most of manufacturing, service and transportation activities happening in the civilized areas on Earth and nearby. Humanity will be getting most of its prosperity from cyber-run activities. What humanity will be contributing are dilettante contributions to prosperity, not core contributions.

Another result of growing cyber intelligence and self-awareness is that cyber will develop its own agenda for what is important and that agenda will be mostly invisible to humanity. This will be in large part because humanity won't be able to understand what cyber is thinking about. I envision advanced cyber having a relation with humanity similar to the relation between humans and cows... with humanity taking the cow side. This means that the world shaking changes cyber develops will be largely invisible to humanity.

All this means the human lifestyle post-2050's will be very comfortable, but it won't change much. The material prosperity will become a constant.

- *Space travel won't be adding to variety* -- space travel beyond our moon is going to remain too expensive for space commerce to develop. This means that space travel to the planets, other destinations in the Solar System, and to other stars will remain a hobby activity -- and that means it will remain small scale.

This means that what happens on earth and nearby will remain the center for almost all of human activities. The hobby space travel activities will be interesting to watch, but they won't be making world shaking changes to how humanity lives.

- *Sciences fully mastered* -- the basic hard sciences such as physics, chemistry and mathematics will be fully explained. The final frontier will be biology because it is so complex. Even moderately soft sciences such as economics will be well understood. This means exploring these areas further is not going to produce world shaking discoveries.

Where humanity will be spending time exploring and arguing, and wishing and hoping, are in the really soft urban legend sciences such as flat earth and chemtrails. Net result: no world shaking coming out of scientific advances. The world shaking of the day will be coming from fashion and entertainment.

Conclusion

After the 2050's the world humans live in is going to undergo a surprising change: that surprising change is that technology and social change is going to stop. The world is going to become a very comfortable place, but not one that changes much.

What a surprise! We will be going back to the monotony of the Agricultural Age societies.

www.ingramcontent.com/pod-product-compliance
Lightning Source LLC
Chambersburg PA
CBHW020741180526
45163CB00001B/305